Gender and the *Koseki* in Contemporary Japan

The Japanese *koseki* system is the legal and social structure keeping record of all Japanese citizens. Determined by the Civil Code and the Koseki Law, for activists challenging it, the *koseki* is also an ideological structure, which has produced patriarchal control through single-surname households.

Based on ethnographic fieldwork in Tokyo, this book engages with issues of gender hierarchy and structural inequality in Japanese society. Studying several decades of feminist activism and critique of the *koseki* system, it analyses the strategies of activists who have creatively circumvented *koseki* rules in order to maintain their natal names in marriage. It examines the case studies of members of the *fūfubessei* (separate surname movement) and the movement to end discrimination against children born out of wedlock, and in so doing this book illuminates the contradictions in current family law and *koseki* practice that have animated a generation of feminists in Japan.

Demonstrating the effect of the *koseki* on family, gender, and national identity, this book will be useful for students and scholars of Cultural Anthropology, Gender Studies, and Japanese Studies in general.

Linda White is Associate Professor and Chair of Japanese Studies at Middlebury College, USA. Trained as a cultural anthropologist, she has conducted ethnographic research with grassroots organizations in the Tokyo area during the past several decades.

Routledge Contemporary Japan Series

66 **Regional Administration in Japan**
Departure from Uniformity
Shunsuke Kimura

67 **Japanese Media at the Beginning of the 21st Century**
Consuming the Past
Katsuyuki Hidaka

68 **Intercultural Communication in Japan**
Theorizing Homogenizing Discourse
Edited by Satoshi Toyosaki and Shinsuke Eguchi

69 **Local Politics and National Policy**
Multi-level Conflicts in Japan and Beyond
Ken Victor Leonard Hijino

70 **Trauma, Dissociation and Re-enactment in Japanese Literature and Film**
David C. Stahl

71 **Rethinking Japanese Studies**
Eurocentrism and the Asia-Pacific Region
Edited by Kaori Okano and Yoshio Sugimoto

72 **Japan's Quest for Stability in Southeast Asia**
Navigating the Turning Points in Postwar Asia
Taizo Miyagi

73 **Gender and the *Koseki* in Contemporary Japan**
Surname, Power, and Privilege
Linda White

For more information about this series, please visit: www.routledge.com/Routledge-Contemporary-Japan-Series/book-series/SE0002

Gender and the *Koseki* in Contemporary Japan

Surname, Power, and Privilege

Linda White

Routledge
Taylor & Francis Group
LONDON AND NEW YORK

First published 2018 by Routledge

2 Park Square, Milton Park, Abingdon, Oxfordshire OX14 4RN
52 Vanderbilt Avenue, New York, NY 10017

Routledge is an imprint of the Taylor & Francis Group, an informa business

First issued in paperback 2019

British Library Cataloguing-in-Publication Data
A catalogue record for this book is available from the British Library

Library of Congress Cataloging-in-Publication Data
A catalog record has been requested for this book

ISBN: 978-1-138-67434-9 (hbk)
ISBN: 978-0-367-42420-6 (pbk)

Typeset in Times New Roman
by Wearset Ltd, Boldon, Tyne and Wear

This book is dedicated to all the activists in Japan working for the right to keep their own names

Contents

Figures

Preface

"The *koseki* is in here, in their (our) heads" (*koseki wa koko desu*, she said, pointing to her own head, Akaishi Chieko 2012 interview).

Akaishi Chieko, a single mother, feminist, and writer whom I had interviewed about her activism back in the 1990s, invited me to the Single Mothers' Forum Christmas party at a local community center near Yotsuya station in central Tokyo at the end of 2012. Chieko had helped me with my earlier research on grassroots feminist activism and I had joined her and other activists at the Fourth World Conference on Women in China in 1995 where we participated in workshops together, distributed information about the situation of single mothers in Japan and celebrated an exciting time for Japanese and global feminism. Chieko knew that I was carrying out research on gender and the *koseki* (household register), and in particular, on feminist activism to change the *koseki* and surmised that the Christmas party would give me a unique opportunity to meet some women who had "interesting experiences with their *koseki*" and might be willing to participate in my research.

The *koseki* is a government registration system that documents kin relationships by family unit. All Japanese citizens are named in a *koseki* and prove citizenship through their registration. The basic unit of the *koseki* system is the household, not the individual. Two sets of laws, the Koseki Law with 138 articles, and additional supplementary provisions, and the Japanese Civil Code with hundreds of articles covering matters from business obligations to the specific details of family law, determine what can and cannot be documented on a *koseki*. Though people may speak of their *koseki* in very personal and almost intimate terms, equating *koseki* with family, "I didn't want to leave my *koseki*" or "I wanted to return to my *koseki*" (as several interviewees said to me as they described their feelings about marriage and what to do after divorce), the system is part of the state's management and control of the population under the purview of the Bureau of Legal Affairs of the Justice Ministry. The government, not the individual, controls the *koseki* and the information it contains.

The activist and writer Tanaka Sumiko, a plaintiff in numerous lawsuits against the government regarding discrimination in the *koseki* and a key informant in my research on feminism and the *koseki*, identifies several prominent problems with the *koseki* system including the *tsuzukigara*, a part of the *koseki* document which ranks all people by birth order and gender; the requirement that each family unit has only one surname, which generally preserves the patriline of the family and mutes the wife's surname and genealogy; and the multiple ways in which the legitimacy of children is demarcated on the *koseki*. Chieko, too, has been critical of the *koseki* and is well versed in its problems, particularly from the viewpoint of single mothers. She knew that I wanted to interview women about their *koseki* experiences, not just those who were grappling with children born out of wedlock or those who had struggled with the *koseki* implications of divorce and single motherhood, but people with a range of different kinds of experiences with the *koseki*. I told her I wanted to see the *koseki* from the perspective of women, both activist and non-activist, to get a sense of *koseki* awareness and *koseki* problems for average women managing the complexities of daily life. Chieko suggested that I bring some of my surveys and join the Christmas party.[1] I will say more about my use of the category of women and the issue of gender in the *koseki* in the coming chapters. These matters are central concerns of my research. Clearly, the category of women is fraught from an analytical standpoint. What a woman is or how the category of women is constructed must be situated within a particular cultural and historical context. Gender inequality comes in many forms and is created and produced through many systems and ideologies. In Japan, the *koseki* system is a central system of gender formation and gender bias. The problems that the activists I study are facing are very much related to their position in the *koseki* and society as women. Gender problems and the activism to address these problems will be examined in the rest of the book as we see how gender itself is constructed within a limited and limiting dichotomous structure built on social values and norms that shape the laws and the structure of the *koseki* system which has developed over several centuries. Gender biases are encoded in the Koseki Law, the Civil Code and in the material structure of the *koseki* document that tracks families in Japan today.[2]

The Christmas party, with a Christmas tree and present exchange for the kids and lots of snacks and drinks for everyone, was a lively event. Each woman paid a small fee to participate in the "party" and in return was given a sack of goodies and presents for mother and children, reminiscent of the old *ko* system used in farming villages, in which money was collected from the group and redistributed to increase overall wealth in a common system of exchange in rural Japan.[3]

During the party, while everyone nibbled *senbei* rice crackers and drank juice and tea, we went around the room and introduced ourselves in the de rigueur "*jikoshokai*," a set component of many parties and meetings in Japan, in which people give a brief self-introduction, providing basic personal information and often explaining their connection to the group or reason for participating. Several of the women started by giving their name and the place they lived, followed by the number of children they had and how long they had been single mothers. Many came from the metropolitan Tokyo area, though one woman had brought her children from a distant province, so hungry was she for the single mothers' gathering, which she told me she attended annually. I learned quickly that among them were young widows, divorced mothers, and never-married mothers. Each talked briefly about her situation. The most common story was of divorce and the struggles of raising children alone while holding down a job.[4] Other stories, like Chieko's, were of women who had never married by choice and had raised children with non-marital male partners or alone.

Chieko invited me to tell the group about my project. As usual, I found it challenging to describe the project and my research questions without revealing my assumptions. I already knew a great deal about the *koseki* and the activism to change laws affecting it, but with this group of women and potential interviewees, I emphasized my interest in their individual experiences with the *koseki* as single mothers. As I came to understand during my research, I knew more about the workings of the *koseki* than many of the people I interviewed. There are just simply too many intricacies of the laws and effects of the laws for even a well-educated citizen to know unless they have a particular need to know. These women had much to teach me about their specific encounters as single mothers, divorced women, and, in some cases, feminists trying to address discrimination in the *koseki*. Chieko herself has written numerous essays about discrimination against single mothers and their children in Japan.[5] Generally speaking, single mothers are more well-versed in the workings of the *koseki* than married mothers, because they have more occasions to represent their families in the public sphere where the *koseki* is required as proof of identity. Furthermore, the fact of their familial "difference" makes it necessary for them to bear various kinds of bias or ill-treatment for themselves and their children as they engage with the state and other social institutions for access to many of the benefits of citizenship or permanent residence in Japan. Single mothers encounter discrimination because of assumptions about households headed by women, assumptions that proper households should be headed by men and the evident "differences" of their *koseki*. So, in my brief introduction I told them that I was in the process of writing about the *koseki* experiences of women from a range of different household types. I explained that their experiences would help me understand the significance of the *koseki* and the ways it operates in society.

In this informal setting, Chieko made it easy for me to describe my research goals and ask for people's help. And sure enough, lots of women had interesting stories to tell. Some completed my survey right there and then and others sent the survey to me after the busy holidays were over. Quite a few handed me their *meishi* (business cards) and several sent Christmas cards to me and offered to be interviewed. I ended up doing extensive interviews with five of the women I met that day. It is worth mentioning that although some people I met during the year of research were dubious of a project examining something so inevitable as the *koseki* – particularly happily married friends and contacts, male and female, for whom the *koseki* may be relatively benign – among single mothers, I had no need to explain the significance of my research. From this group came comments such as "*yokatta desu, daiji na mon desu*" ("That's good. It is important") and "*Naruhodo, koseki da ne*" ("It makes sense, the *koseki*"), suggesting understanding and hope that the logic and significance of the inscrutable *koseki* might be analyzed and questioned. They immediately understood why the *koseki* was worth studying.[6]

Still, in some ways it was a conservative group. The women were just trying to give their kids a "fun Christmas party where *sekken* (the eyes of society) would not hover in judgment over their families," families generally viewed by Japanese society as inadequate because of the lack of husbands/fathers. We sang a few Christmas carols, opened gifts that had been exchanged in a secret Santa kind of game, and the kids did a little skit. There were no radical proposals for group childrearing, same-sex marriage, or mounting challenges to Japanese patriarchy. Rather, the women talked about trying to give their children more time, resources, and love despite being single mothers. Overwhelmingly, this was a social event. At the same time, the social event for families headed by women, a single mothers' Christmas party, had sprung from the societal condemnation of such families within the sphere of taken-for-granted compulsory heteronormative family culture in Japan.

Chieko, along with several other members of the Single Mothers' Forum, a grassroots organization that promotes the interests of single mothers through lobbying for legal protections, published a book in 1994 called *Cheers to Single Mother Families* (*Boshikatei ni kanpai*), refuting the idea that children born in single-mother households were somehow to be pitied by society. Creating social events designed for single-mother families was intended to make kids feel more "normal" and less ostracized and to provide an environment of acceptance and support, Chieko told me. Many of the women talked about how happy they were to be with old friends who understood the realities of their busy lives.[7]

When I interviewed Chieko in October 2012 about her work with single mothers and gender issues, she told me that there have been fewer institutional requests for the *koseki* in recent years than previously. She told me that the dilemma that single mothers used to face, being constantly forced to show their *koseki* and declare their *boshikatei* (mother-child or single- mother household) status, has changed with new privacy laws passed in the late 1990s. The enactment of the privacy laws limits non-kin members from accessing a person's *koseki*. In the past, lack of privacy regarding family matters was a central problem with the *koseki* system in modern Japan, where divorce, out-of-wedlock children, and other personal family statuses were cause for discrimination.[8] Among the problems faced by single mothers in Japan today is the stigma of being single mothers in the first place, which is glaringly clear each time the *koseki* document is required to carry out public business, such as registering for school, applying for a job, or signing a lease. Still, Chieko said, pointing to her own head, "people have the *koseki* inside here, that is the problem." Her embodied reference to *koseki* ideology, an "inside-their head" thing, reveals the at-once ephemeral and material nature of the *koseki*. For many people who have shared their stories with me through the years, the *koseki* is almost equivalent with feelings of family. And yet, as a document whose content is rigorously stipulated by law and thus controlled by the state, it may misrepresent or create problems for their families.

The identity derived from one's position within a family, citizenship status, and gender are all specified on the *koseki*. People's understanding of their social location is often derived from the *koseki*, as *hittōsha* (first registrant), *chōnan* (eldest son), *chōjo* (eldest daughter), and so forth. People I interviewed make sense of who they are through the terms and positions that originate on the *koseki* to designate family position. Such positioning may start with *koseki* categories and reverberate across various public and private spaces because the *koseki* is an essential source of proof of these identity markers for Japanese people. Related to this is the tendency to refer to people by their status identifier rather than by name. Many wives are referred to respectfully as *okusan* rather than by their given names of Sachiko, Mari, or Masako. Non-use of first names within the family is very common. Even neighbors and friends are likely to use *oyomesan*, the term for daughter-in-law, even though she is not *their* daughter-in-law. A husband might be referred to as *otto*, *shujin*, or *goshujin*, showing a range of subtle differences in tone and level of formality, respect, and attitudes about gender, and both a mother and father may be referred to by outsiders as *otōsan* (father) and *okāsan* (mother).

Over the course of 11 months preceding and following the Single Mothers' Christmas party I gathered *koseki* stories from old friends and new contacts, primarily in the Tokyo area but also in Kobe, Yokohama, and several other places around Japan. In addition to laments of inequality and discrimination, I talked with people who believed the *koseki* system plays an important and unproblematic role in Japanese society. And over the year of intensive research and since then, I've had many more conversations with people in Japan and the United States about the fundamental meanings and effects of the *koseki* system. The acquaintances, friends, and interviewees I have talked with and interviewed understand that the *koseki* has a power beyond the paper it is written on in shaping the ways in which Japanese citizens understand themselves and their relationships in their families and in society. But there is no simple agreement about what is most troubling about this complex registration system that locates each citizen within a family and the nation, and, thus, within the world.

Notes

1 My ethnographic research included a survey that was returned from 160 adults and another 75 students.

2 Molony and Uno provide a helpful discussion of gender and biology in historical analyses:

> [w]e do not believe that gender is an essential derivative of "biological" sex—indeed, "sex" is given definition through "gender"—but we do believe that laws, customs, ideologies, and the structures of societies, based as they have often been on the idea that gender is biologically determined, contribute to the social construction of gender.
>
> (Molony and Uno 2005: 2–3)

Indeed, anthropologists have emphasized since Margaret Mead's famous study in Papau New Guinea that gender varies by culture, and that it is precisely this variance that allows us to view the constructed nature of systems of gender (Ortner 1974). In her important essay, Kano (2005) reminds us to avoid using an analytical "transhistorical femininity" assuming that gender categories are somehow stable through time. Butler (2004) provides important rubrics for understanding the performative and instantiative aspects of gender in practice, linked only in our minds to biological variations. See McLelland (2000) for an excellent discussion of gender and sexuality variance in Japanese society. I am grateful to an anonymous reader of an earlier draft of the manuscript who pointed out the need to elaborate my understandings of the category of gender and my use of the category of "women."

3 Smith and Wiswell (1982) discuss this exchange network that was used in Suye Mura (an eponymous village in southern Japan and the subject of the book of the same title by John F. Embree) as a way to raise money and essentially create loans for members of the group. All members contributed a small amount of money every month and each month one member took the whole pot. In the case of the Single Mothers' party, there was no need to charge for the event since everyone brought food and drinks and contributed them. However, by asking for a contribution, the organizers were able to buy high-value gifts and foods to redistribute back to members so that their 1,000 yen bought them things worth approximately 5,000 yen (1,000 yen, equal to US$8.50 in 2012).

4 Many divorced women I interviewed over several decades told me that they did not receive alimony or child support from their ex-husbands. There are many reasons for this, one of which is some women's preference to break all ties with an ex-husband. See Hertog (2009) for further discussion of the low rate of child support paid by ex-husbands and biological fathers who are often married to other women in the case of out-of-wedlock births.

5 Akaishi has written many essays and pamphlets in Japanese discussing the economic and social disadvantages that single mothers face in Japan as outsiders to normative heterosexual marriage. For an English translation of her work see "Single Mothers" (2011).

6 In other contexts in Japan I get almost blank stares when I tell people what I am studying. For many Japanese, the *koseki* is so taken for granted and such a straightforward record of family that it doesn't seem worthy of research. See Chapman and Krogness (2014) for a reference of this same misapprehension among some of the people they interviewed about the *koseki*.

7 The Single Mothers' Forum holds numerous annual holiday events like the Christmas party, including a Cherry Blossom Viewing party in the spring and a party during summer vacation. These outings are scheduled at times when not having a normative nuclear family can be particularly difficult for children or their single mothers.

8 Ninomiya writes of the easy access to others' *koseki* from the late nineteenth century that came to characterize the concern and anxiety people felt about the information within their *koseki*.

> It became easy to examine *koseki* ledgers and request full or partial copies of *koseki* documents. As a consequence, discrimination based on family background and prejudice targeting, for example, divorce and extramarital children emerged. We see a growing consciousness of the legitimate family (seitō na kazoku ishiki) and the emergence of a concern for maintaining this legitimacy and a desire to regulate conduct to avoid such deviation.
>
> (Toshitani 1987: 140, cited in Ninomiya 2014: 175)

Acknowledgments

Writing a book about the *koseki* has taken a long time and has been possible because of extraordinary help and encouragement from many people along the way. I would like to thank my key informants in Japan, good friends and determined activists working on multiple fronts, often holding down full-time jobs, caring for their families, and actively participating in feminist and other grassroots organizations, who made time to help me. Barbara Summerhawk, American by birth and deeply embedded in her adopted land of Japan, first introduced me to the *koseki* and *koseki* activists in 1994. She has been an amazing friend, aikido teacher, and mentor for over 30 years. Thank you to Yumiko Jansson-Yanagisawa for sharing the joys of feminist work, literature, and life, and always encouraging and enriching my inquiries about gender and feminism in Japan. We, too, have had a long, more than 30-year, friendship. Special thanks to Mikiko Ishihara for many years of friendship and generous support of my work and research in Japan. These three exceptional activists, writers, and translators have introduced me to many women whose stories and experiences have shaped this book.

Over the years, many people have talked to me about their personal experiences navigating complex lives to bring about social change in Japan, but none so much as Tanaka Sumiko who has been a generous friend and a non-stop source of information about the *koseki*. She has provided much of the first-hand data on the legal cases that have led to reform of the *koseki* system in the past 30 years. Her extraordinary mastery of the complex laws, practices, and norms of the *koseki* system and the lawsuits against the government have anchored my research. Her breath-taking commitment to getting this all down on paper (in Japanese) has left a paper trail of each lawsuit, each lecture given to educate community members about the *koseki*, and each step of a multi-decade campaign to raise awareness about the gender bias embedded in the *koseki* system. The only way to even partially repay her for all her generous help is to put some of this material into English so that it might reach a wider audience. Many other activists, lawyers, and participants in lawsuits shared their expertise and time to help me along the way. I have changed the names of many of my informants to protect their personal stories and privacy, but without their generous participation in my research I could not have written this book. I would like to thank by name several key sources who generously explained important aspects of the *koseki* system to me during my research. I offer my sincere gratitude to Professor Sakakibara Fujiko of Waseda Law School and the Women's Law Practice, Sakamoto Yoko of the Besseisoshō o sasaeru kai (Organization to Support the Bessei Lawsuit), Harumi Sakai, Oguni Kaoru, Akaishi Chieko (Single Mothers' Forum), and Tomizawa Yoshiko. And many thanks to good friends Saeko Jozuka who worked as research assistant during 2012 and 2013 and Chikako Kobayashi for her ongoing help with my research.

Professor Inoue Teruko, my first academic mentor in Japan, helped me understand the historical salience of family law and the development of women's studies in Japanese institutions of higher education. We spent many wonderful lunches together in preparation for the Fourth World Conference on Women in 1995 and she has continued to support my work through invaluable introductions and conversations. Thank you to Professor Kuniko Fujimura-Fanselow for her enthusiastic support of my work and for sharing her own *koseki* stories. Professor Reiko Yoshihara, feminist and activist, generously welcomed me as a visitor at Nihon University during the 2012–2013 research. Thank you for providing an academic home.

A generous grant from the Japan Foundation funded my research on Japanese feminism in the 1990s. Many of the ideas that fostered my interest in the *koseki* first developed in Japan in 1994–1995 while a graduate student at the University of Colorado at Boulder. I am most grateful for the mentoring and friendship of my Boulder community of anthropologists and feminists, Anne Allison, Charlie Piot, Dennis McGilvray, Paul Shankman, and Marcia Yonemoto, for guiding my inquiries into Japanese feminism and family law. Susan Schmidt, Lisa Park, Rachel Silvey, David Pellow, and Michiko Hase, each in her/his own way, provided precious feedback and support making the work we did together fun and exhilarating.

Thank you to mentors and guides in Japanese anthropology, Glenda Roberts, Susan Long, Jan Bardsley, and Amy Borovoy, for helpful conversations and recommendations about my research on Japanese feminism and the *koseki* issue. I am very grateful to the Fulbright program for research funding in 2012–2013 to carry out the interviews and ethnographic research upon which the book is based.

At Middlebury I have many people to thank. I am grateful to the Professional Development Fund and Dean Jim Ralph for summer funding to return to Japan several times to meet with contacts and update information for the book. My colleagues in Japanese Studies, Carole Cavanaugh, Nobuo Ogawa, Steve Snyder, Masa Takahashi, Sayaka Abe, and Kyoko Hayasaka Davis have patiently listened to versions of my research and supported me in many ways along the way. In particular, Kyoko Davis has answered many an errant question about the reading of a character, or the meaning of some obscure comment in my interview material. I can never adequately thank you all. And many thanks to Judy Olinick, Coordinator for Japanese Studies, for office and moral support throughout the years. I am grateful to everyone in Gender, Sexuality, and Feminist

Studies for encouragement and support, and in particular I want to thank Sujata Moorti without whose encouragement and faith this project might have never seen the light of day. William Poulin-Deltour has been a wonderful friend since our first meeting upon our arrival in Vermont. Thank you to mentors and friends who have taken a great interest in my work and reassured me when I was down, especially Natalie Eppelsheimer for juice and bootcamp, and Peggy Nelson and Barbara Hofer for their generous feedback and encouragement in the writing process. I am also grateful to the wonderful Middlebury undergraduates who have served as research assistants over the years, especially Peter Moore and Mari Tanioka. I would also like to thank two anonymous reviewers who provided extraordinarily helpful comments that guided my reworking of several key points of the book. I will always be most grateful for their time and work.

My thanks and love to John and Emma Snyder-White who make everything I do worthwhile. Emma's support as junior research assistant in 2012–2013 in Tokyo and again in the final weeks of book layout during her summer break from college has shown me the power of one girl. Steve Snyder helped me think through many of the interesting problems of family and power in Japan as the manuscript developed and helped at all stages of this project. Thank you, Steve, for keeping me sane and not letting me give up. You have played such a fundamental role in this work and in all aspects of my happy life.

During the last several years of this project my sister, Chris Booth, lawyer and editor par excellence, provided very helpful feedback and editing with the eye of a legal insider and Japanese cultural outsider, showing me where things I had taken for granted needed to be spelled out more thoroughly. During the very last months of my work, Carolyn Sachs, cousin and fellow ethnographer, curious about my project, joined in with very productive comments on a late draft. Thank you both!

Finally, thank you to my generous editor Stephanie Rogers and her thorough and patient assistant Georgina Bishop at Routledge for support and guidance in this long process. It has been a pleasure to work with both of you.

Introduction and chapter overviews

In 2014, Routledge Studies in the Modern History of Asia produced the first book in English to examine the Japanese household register, the *koseki*.[1] In the introduction to that work, the editors David Chapman and Karl Jakob Krogness describe the *koseki* system as one embedded in, and productive of, the nation-state. They describe a system concerned with "population governance, social control, and the maintenance of social order" that served the rapidly modernizing Japanese state from the earliest days of the Meiji period (1868–1912).[2] For them, "this extremely powerful state instrument" provides insight into both historical and contemporary understandings of citizenship, the family, and the relationship between citizens and the state. Individual chapters of the anthology address particular historical periods, before and after the modern *koseki* was implemented in 1871, and particular groups of people, including women, whose experiences in Japan prior to, and since, 1871 have been mediated through the *koseki* in unique and complex ways. Because the *koseki* system catalogs citizens, it provides an important lens into understandings of insiders and outsiders, citizens and others, men and women, across time. One-third of the anthology is dedicated to contemporary gender issues related to the *koseki*, including a chapter I authored entitled "Challenging the Heteronormative Family in the *Koseki*: Surname, Legitimacy, and Unmarried Mothers."

This book builds on the discussion of gender in the *koseki* begun in the Chapman and Krogness anthology. Although citizenship status, ethnic identity, and family heritage are all keen concerns in any study of the *koseki*, I focus this study of the *koseki* on feminist analysis and critique of this basic Japanese identification and census system. The book is based on my ethnographic research in Tokyo, focused primarily, but not exclusively, on feminists challenging the *koseki* system. Through interviews, observations, and informal conversations I have gathered a range of *koseki* stories to illuminate how the *koseki* defines and shapes gender and informs people's stories of their lives.

My goal in doing ethnographic research on this topic was to further understand the complex *koseki* system itself, and to deepen my understandings of the problems women face as they come to experience their identity in the *koseki* as daughters, wives, daughters-in-law, mothers, and so forth. Through the process of collecting stories about and hearing critiques of the registration system, I came to understand the gendering nature of the *koseki*; how it reflects gender patterns, norms, and expectations, and how it simultaneously produces gender and understandings of gender in Japan. As Judith Butler argues, gender is a continuously produced set of meanings that demands and specifies particular parameters for a person's behavior and comportment so that they will be intelligible to others as either male or female in a gendered society (Butler 1990). The *koseki* provides one more site, and arguably a very central site, within Japanese society for the *iteration* of gender. But its particular power and effects result from the way in which it affixes gendered identity or gendered subjectivity to other identity statuses located within the family and the nation. At the level of the *koseki* this gendering has implications for all facets of society, since the *koseki* is an essential tool and naming instrument with which Japanese claim subjective identity and navigate social relations and institutions.

Methods and objectives

Over the course of multiple year-long stays in Japan since the early 1980s I have been studying feminist activism and changing gender norms in Japanese society. From August 2012 through July 2013, with several additional weeks in June 2012 and May and June 2014, primarily in the Tokyo area, I followed the introductions of my friends and activist contacts to meetings, lectures, court hearings, and other events, including demonstrations and parties, addressing a wide range of feminist issues and, in particular, *koseki* issues. During this period of ethnographic fieldwork in 2012–2013, in addition to many unrecorded conversations and informal interviews, I carried out 35 in-depth, digitally recorded formal interviews. Six of the 35 interviewees were *koseki* experts, three civil servants who had experience working in the *koseki* division of their government offices and three lawyers who had worked on *koseki*-related lawsuits.

The non-expert interviewees varied greatly in their knowledge and experience with the *koseki* system. Interview length depended on each individual's experience and eagerness to talk. I usually did a single in-depth interview ranging from about one-and-a-half to four hours. However, with some of the most politically actively engaged interviewees I did multiple recorded interviews over a period of months. Several of my interviewees, like Tanaka Sumiko, have helped me with this research for years. Tanaka knows as much as anyone about how the *koseki* works because of her own personal study of the *koseki* and her involvement in lawsuits for several decades. I first met key activists, Tanaka Sumiko, Akaishi Chieko, and Ishihara Mikiko, in 1994 and they have generously contributed to my work and understandings of changing family structures, the *koseki* system, and a wide range of other feminist matters over a long period of time.

For the most part, the interviews were open-ended and tailored to the experiences of each participant. I asked both female and male interviewees to simply recount their experiences with the *koseki*. The *koseki* was often a starting point for discussions of family, adoption, divorce, and other personal matters. The youngest person I interviewed was in her twenties and the eldest was over 70, but many of the people I interviewed were in their forties and fifties. The interviews with non-experts were extremely rewarding. I was able to listen to the narratives of people's lives, often defined by their key relationships and sometimes their political involvement. These interviews tended to tack between personal (and often painful) experiences and lighter, more conversational anecdotes about family history or personal interests, such as music, gardening, and travel. Of the non-experts, 24 were women, three were men, and two were transgender. Of the 24 women, one was Zainichi-Korean and one was American; the rest were in *koseki* as Japanese citizens.[3] Three identified as currently in lesbian relationships and three had been married or still were married to foreign men. Of the three non-expert men, two were Japanese and one was American. All three men were in relationships with Japanese women. Interviewing men about their experience with the *koseki* informed my understandings of the differential effects of the *koseki* on men and women. I did not seek male interviewees, but gladly interviewed them when they volunteered to participate in the research.

Ethnographic research has become increasingly unbounded by geographical place because of the Internet. Now, when I leave Japan or even leave a particular interview, I know I have some access back to individuals and information in ways that were nearly unobtainable when I started living in Japan in the early 1980s. The first time I said goodbye to good friends as I left Japan after living there for three years in 1985, I feared I would never see some of them again.[4] Email and the web make it easier to stay in touch, confirm information, and keep up with ongoing activities in Japan. But despite the ease afforded by new communication technologies, there is still nothing quite like a face-to-face interview. I love the intimacy of such encounters and I cherish the stories that people share with me. These stories form the foundation of my understandings of the *koseki*, my research, and this book.

As a supplement to the qualitative side of my research I distributed a questionnaire through networks of friends, activists, and acquaintances to gain access to more people and a wider range of attitudes about the *koseki* and people's experiences with the *koseki*. The development of my survey questions with two key interviewees and informants, Saeko and Mikiko, was instructive. We went back and forth for nearly a month refining the questions and simplifying the form until the survey barely looked like the open-ended qualitative document I had started with. Generally, my friends suggested deleting most of the open-ended questions, as they were "too hard to answer" (*kotaenai deshō*) and "too vague" (*aimai deshoō*). In the end, the survey I used was not entirely satisfactory, but its development provided insights into what kinds of questions could be asked or answered and helped me find ways of wording delicate questions. In addition, the surveys provide confirmation of the prevalence of certain kinds of discriminatory practices that were explicated in individual interviews. Ultimately, the one-on-one interviews and informal conversations turned out to be more effective for accessing emic understandings of identity, *koseki*, and gender, and my analysis depends more on these components of the research.

In the book I use the information I gathered through interviews, meetings, surveys, and a wide range of pamphlets, newsletters, and other print material to capture women's, and occasionally men's, experiences with, and trouble related to, their *koseki*, as they go through life changes that may require a new name or other revisions to their *koseki* document and/or identities. Not only women are discriminated against in the *koseki* system. That is one reason I was open to interviewing men who had come to question the *koseki* and its influence in their lives. Like gender status, citizenship status, "*burakumin*" (outcaste) status, and sexual orientation all have a way of "showing up" or being found out through notation or some type of marginalization on the *koseki*.[5]

The *koseki* document is used to ascertain the quality and content of another person's family. People care about the appearance of the *koseki*, their own and others'. Marriage negotiations are intricately linked to the other party's *koseki*. Not besmirching one's *koseki* becomes a central motivation to avoid divorce, get married before a child is born, or marry one child ahead of another. When looking at a *koseki* people ask, are there divorces, unmarried adult children, adoptions, or other irregularities that might be used to impugn a family's reputation? Information on the *koseki* is the basis for others to determine if a family is a "good" family or a "bad" family. People who were not associated with *burakumin* (outcaste) community, for example, would try to make sure that marital partners were not of *burakumin* descent by obtaining a copy of the other family's *koseki*. Along these lines, historically, non-marriage of an elder son or daughter or absence of grandchildren might have been perceived as equally problematic as divorce. There is a vast amount of information contained in a *koseki* and multiple planes of identity revealed there. I've chosen to focus primarily, but not exclusively, on gender; examining the male privilege and female subordination that occur through the system of registration that defines Japanese citizenship. Within the frame of gender I might have focused on a particular group of men or women, only divorced women or only married men, but instead I followed the lead of activists in the Tokyo area who have various *koseki* experiences and household forms. The most common attribute of the people I interviewed and studied may be their shared sense that discrimination in the *koseki* affects women deeply and structures women's second-class citizenship in Japan today.

Many of my interviewees had close contact with or were in relationships now or in the past with individuals "othered" in the *koseki* because of their citizenship status, sexuality, adoption, or illegitimacy status. In this sense, as I try to gain an understanding of the experiences of a varied and complex set of Japanese women, I do not perceive them as a simple and homogeneous group. Despite a myth to the contrary, Japan is a heterogeneous society in many respects, including socio-economic class, ethnicity, sexuality, and increasingly gender, given the legal status of transgender people.[6] The *koseki* document exposes many of those identities. As a tool of identification it also serves as a tool for differentiation and ultimately discrimination.[7]

Although the *koseki* is designed to track and distinguish citizens from non-citizens, the very notion of citizen has changed through time.[8] This is one of

Krogness's key points when he argues that citizenship in Japan has been based on, not only reflected in, the *koseki*. *Jus koseki* (or "rights derived from" *koseki*), he argues, rather than *jus sanguinis*, is the law of citizenship in Japan, because, more than anything else, by virtue of being named in the *koseki* one is understood as having Japanese citizenship.[9] Non-citizens, such as foreign spouses, can be noted on a Japanese citizen's *koseki*, but in the margin, not in the main position for a spouse, making their non-citizenship and marginal position obvious. Similarly, people without Japanese blood or heritage (*sanguinis*), such as Caucasian Ogasawaran islanders and Korean subjects of the Japanese empire during the first part of the twentieth century, have been granted citizenship and have been registered in *koseki* when it served the state's interest to consider them citizens and register them. In other words, the apparently obvious status of citizens, those with Japanese parentage, has not always held since Japan defined citizenship in the early Meiji era. As will be discussed later, children of Japanese parentage whose parents do not register their birth in a *koseki* lack citizenship.

With this in mind, we might say that if the citizenship of a Japanese male is the standard of citizenship, then Japanese women through a range of material differences made through both law and custom have a kind of "partial citizenship" conferred through the *koseki* law and made tangible on the *koseki* document. Whether this legally specified partial citizenship "reflects" their second-class position in the society or is a key source of partial citizenship is a question that runs through this book.[10] Activists now fighting for greater equality in the *koseki* at the level of categories of information and structure of the family unit on the document, which is centered on a single address and surname that define the unique household, highlight not only the latent male bias in the *koseki* document and terminology, but the broader societal effects of that bias.

The *koseki* stories I have gathered and examined in this project expose fundamental aspects of male privilege embedded in the *koseki* system, and reveal the "partial citizenship" of even privileged Japanese women. Women's stories of their marriages and divorces, and their name changes, shape not only their understandings of the *koseki* but also their understandings of what it means to be Japanese women today. Everyone is gendered in the *koseki*, as male or female, first son or first daughter, and so forth, but the bifurcated gender designations also produce tight categories of identity that shape consciousness about what a male or female can be.[11] Thus, the book builds on my interviewees' lived experiences, their subjective identities, and their experiences with their *koseki* to reveal a deep level of gendering that perhaps shapes other cultural practices and norms that have received more study in Japanese studies, such as gender and education, gender and sexuality, and the nexus of women's work and family life in Japanese society.

Chapter overviews

In Chapter 1, "The matter of names and why names matter," I examine the development of family names and family identity historically as a way to more deeply understand the surname debate at the heart of the *koseki* problem in contemporary society. In particular, the linkage between geographical place, social status, and surname established in ancient times and through the medieval period among elites shows the tight association of people with their kin-group identity, an identity that congeals in family name in the modern *koseki*. This examination of naming practices as well as name-changing practices prior to the Meiji period is the basis for claiming that names *do* matter for women despite the hegemonic idea that women "should" give up their names in marriage. Through interview and survey material I examine some of the emic understandings of family and identity through the symbolic structure of names. What are the implications of male-named family units? What are the implications of name changing on women's lives? Material from interviews with women who struggle to find an identity once they have left their natal *koseki* helps us understand what it means to have a non-permanent name.

In Chapter 2, "Separate-surname activism," based on interviews and ethnographic material, I examine the multi-decade movement for separate surnames in marriage. This central feminist campaign to alter one of the most basic aspects of the *koseki*, namely, a requirement that a family has only one surname, shows some of the key feminist actors at work as lawyers, writers, and plaintiffs, and lays bare the structural bias in the modern (since 1871) *koseki* system, constituted fundamentally around the single-surname rule.

In Chapter 3, "Common-law marriage as a form of *koseki* resistance," I discuss the experiences of those who opt out of the normative *koseki* registration regime in order to resist "patrilineal and patriarchal" aspects of the *koseki*. Numerous activists I interviewed are in what in Japanese is called *jijitsukon*, translated as common-law marriages. Such marriages, which are essentially unregulated by law and thus have no rendering in the *koseki*, serve as a profound rejection of the *koseki* and normative family structure. *Jijitsukon* are legally invisible. The partners are in separate *koseki* and the legal invisibility of their relationships leads to a number of consequences that inevitably result from non-marriage, including the apparent "illegitimacy" of children. Several of the most prominent *koseki* feminists have zeroed in on the bias toward unmarried women and their children by suing the government for discrimination against children. Key common-law marriage activists and several lawsuits claiming damages due to discrimination against children and unmarried women in the *koseki* will be discussed.

In Chapter 4, "Illegitimacy and male privilege: the underlying logic of the *koseki*," I examine the idea of illegitimacy and its formidable role in the *koseki* system. I trace understandings of male sexuality and kinship from the Meiji period to the present to better illuminate changing understandings of male sexuality and parental responsibility in non-marital offspring. There is a deep logic that treats children from non-marital relationships as not blood-related, although they are the father's children. What is the meaning of the distinctions between legitimacy and acknowledgment as they are inscribed in the *koseki* structure today? How do these categories of personhood and kin relationship inform gender in the *koseki*?

In Chapter 5, "Beyond the scope of the *koseki*: families out of bounds," I identify some of the starkest contradictions and unlikely ironies in the *koseki* system today. Despite the elaborate governmental bureaucratic systems designed to keep track of every citizen and make sure that all notifications in registrations are accurate, there are a number of kinds of inaccuracies that proliferate precisely because of the discriminatory and highly heteronormative structure of the Civil Code and *koseki* system. These misrepresentations include: the exclusion of non-Japanese marital partners from the spousal column on the *koseki*; same-sex partner adoption (a strategy for marriage where marriage is not possible); non-registration of children who are clearly Japanese by ethnicity and citizenship of parents; and "illegitimate" children who are living with and being raised by their unmarried parents in common-law marriages. These misrepresentations on the *koseki* document expose deep contradictions in a system that is increasingly out of step with the fluid and changing nature of Japanese kin groups, where divorce, re-marriage, out-of-wedlock childbirth, in-house divorce, and non-marriage are increasingly common occurrences. Indeed, these mis-registrations expose an uneasy relationship between past norms and current practices. The Meiji-era logic of the *koseki* structure, a male-headed household with marrying in wives and marrying out children, bleeds through into the twenty-first century *koseki* document, revealing the displacements or mis-registration of people and family members. Displacement, contradiction, and misrepresentation result from the clash between the rigidity of the *koseki* system and the fluidity of always changing human relationships.

This observation is parallel to, but different from, the important observation made by Chapman (2014) that the 1871 Koseki Law, which was implemented to create social order from social chaos through classification and documentation, in fact produced a new kind of chaos, as people did not fit the simple categories of the *koseki*. Here, I argue that because the *koseki* is a basis of discrimination in contemporary society, some people intentionally mis-register their identities in order to escape discriminatory designations for themselves or their family members or, as in the case of same-sex partners, in order to gain rights through a fictitious parent–child relationship. These registration strategies may make kin relations possible or less discriminatory but they lead to inaccuracy and a kind of chaos in terms of "grasping" the population accurately (Chapman and Krogness 2014). This intentional mis-registration is an important part of feminist resistance to oppressive aspects of the *koseki*.

Let us turn then to the central issue of surnames in the study of the *koseki* system. Universal surname usage in Japan corresponds with the modern *koseki* system established through the Koseki Law in 1871. Surnames, like permanent addresses (*honseki*), organize the *koseki* document and the fundamental understanding of family. In Chapter 1 we examine the history of the development of surnames and the *koseki*, beginning with the story of Keiko Nakayama who lost her maiden name and then worked to get it back.

Notes

1 One of the important contributions of *Japan's Household Registration System and Citizenship: Koseki, Identification and Documentation* is the variety of perspectives on the *koseki* that it provides. Half of the essays were written in Japanese and translated for the volume. Writers in English come from Europe, Australia, North America, and Asia.

2 Chapman and Krogness (2014: 1).

3 Many Zainichi-Korean self-define themselves with this term to represent their complex position within Japanese society. Most people so defined have not naturalized and do not have Japanese citizenship despite being second or third generation born in Japan. For a good discussion of this identity group and the history of Koreans in Japan see Sonia Ryang (2000), David Chapman (2008), or John Lie (2008).

4 I am extremely grateful to many people who have helped me return to Japan many times since 1985 to do research and live in western Tokyo. Besides teachers and mentors who encouraged my research, I would like to thank the Japan Foundation, SSRC, and Middlebury College professional development fund for generous funding that informs this project. In 2012–2013 a Fulbright Research Grant supported the *koseki*-dedicated ethnographic research for this book.

5 *Burakumin* or *hisabetsu burakumin* refers to a particular group, identified, not so much by ethnicity or any other shared traits as by a shared experience of discrimination based on heritage. See Amos (2014) for a discussion of the *koseki* and this group during the Edo period and Ian Neary for a discussion of *burakumin* identity today.

6 Gender Identity Disorder was acknowledged in Japan and became the basis of gender change in the *koseki* in 2007. See Ninomiya (2014) for a thorough discussion of some of the problems with that law.

7 See Lie (2008), Chapman (2012), and Ryang (2000) for valuable studies of the marginalization of people lacking Japanese citizenship in Japan. Although non-citizens are simply excluded from the *koseki*, other minority groups, particularly *burakumin*, people of Japanese ethnicity who may be passing as "regular" citizens are revealed (exposed) through the *koseki*, and thus the *koseki* becomes one of the tools for their oppression.

8 See Kamoto (2014) and Chapman (2014).

9 Krogness and Chapman (2014) examine the contradictions between Nationality Law and *koseki* practice. Unregistered children and children born to unregistered mothers lack Japanese nationality because they lack registration.

10 See Wildman (2000) for an excellent discussion of the workings of male privilege.

11 Butler argues that gender is produced through its iterations and the *koseki* is a kind of ground zero for gender specification, it is simultaneously a birth certificate, a family tree, and the primary source of identification for Japanese of all ages, by and within families.

1 The matter of names and why names matter

Nakayama Keiko met me at a lunch spot near Mitaka station, not far from my Kichijoji apartment, on a crisp fall day in 2012 to discuss her *koseki* and a particular set of family problems that had led her to change her name in the past few years. When Tanaka Sumiko, a key informant, had suggested that I contact Keiko for an interview, she had referred to her as Koyasu-san, not Nakayama-san. However, as we set up the interview it became clear to me that her surname was Nakayama, and as she told me her story of marriage, divorce, and name-changing, I learned why she had changed her name and why there was so much at stake for her regarding her surname. While I addressed her as Nakayama-san in our interview because of the normative use of surnames in such settings in Japan, I will refer to her here as Keiko. The practice of using the surname rather than the given name in many business and formal settings in Japan further reveals the importance of surnames.[1] I have found great variation in use of surname and given name in my experiences in Japan. Terms of address vary somewhat by context and by generation, by region of the country and by family custom, but one can assume until they are told otherwise, that using a surname, as in Suzuki-san, White-san, and so forth, is the most polite and proper thing to do. This practice reveals the linkage between identity and surname (family name) that is broken when women change their names in marriage.

During our interview, the 61-year-old Keiko told me that she came to Tokyo for college in the 1960s, eager to flee the conservative ways of her small hometown in central Honshu, not far from Mt. Fuji, in Shizuoka Prefecture. After discussing the weather, the coffee, and my research agenda, I asked Keiko about her experience with the *koseki*. She took a deep breath and then launched in. "I had my first child when I was still a student," she said. "I was only 21." "I changed to Koyasu. I was happy to do it." "They say, marriage is a woman's happiness" ("*Josei no shiawase da to*"). "After we had the second child and things got bad, I left him." "I took the boys and left him" ("*tobidashita*"). This torrent of information came matter-of-factly from the petite and attractive Keiko.

She had married her college boyfriend and dropped out of school when she became pregnant. After her divorce she raised the children alone as a single mother in Tokyo. Keiko told me that although she had wanted to take her maiden name back after the divorce, she was discouraged from doing so. "I am from a small town," she said. "It is really old-fashioned" ("*inaka desu yo*"). "People are sensitive about these things" ("*Kodawari ga arun desu yo. Dakara, namae mo, modosu nan te, oya, to ka, kyodai ni meiwaku da to iwarete*"). "So, I was told that it would create a fuss (cause problems) for my parents and sister if I were to change my name. So I kept his name, Koyasu." Changing her name back to Nakagawa would have made it clear to family members and members of her small town that she was divorced. Without the name change, no one in the hometown would need to know, she said. This was not the first time that I had been told that changing one's name after divorce would create problems for other family members and the family's overall reputation. For some of my informants, a shared marital name seems to prove to the world that a couple is still married, regardless of evidence to the contrary. In an unfortunate and unlikely set of events that pre-empted the following tale, soon after Keiko's parents died in old age, her sister who had lived near her parents died prematurely. Thus, within a short number of years all of Keiko's immediate natal family members were gone. As a result, Keiko found herself the sole provider of memorialization rituals for her deceased parents and sister (as well as for the ancestors) back in her rural hometown.

"I was taking care of the family grave in Shizuoka, going there four times a year," she said. "If I had not done it, no one would have." "One time I asked the head priest at the temple if I could be buried with my family." "I said, *namae ga chigau kedo ohaka ni hairitai* (I know my name is different, but I'd like to be buried here – enter the grave)." "But he told me that because my name was different, I could not be included in the family grave" ("*namae ga chigau kara dame*"). "*Sabishii*," she said, "it is lonely, or it makes me feel sad/lonely."

> I return to take care of the family grave in the countryside four times a year, but if my name is not their last name, I cannot be buried with them. That is why I changed my name back and *returned to the koseki* last year (*dakara kyonen, koseki ni modoshita*).

Here we see the interconnected constellation of surname, *koseki*, lineage, and sense of family belonging, which entails memorialization of the ancestors, that seem to define Keiko's social location and family membership. She says she felt *sabishii* for being closed out of her natal family, all the while fulfilling her filial duty to her parents, sister, and ancestors as if she were still a member of that household. Through the process of "returning" to the family *koseki* and family name she was able to reunite with her family and also anticipate joining them at the temple where their ashes are maintained and revered through regular rituals when she dies. Keiko's story shows the overlapping spheres of identity, religious/ritual practice, and emotional well-being connected through the *koseki* document and the name upon which it is centered. Through the legal process of changing her name and re-entering her family *koseki* she has been able to reclaim family membership in the present and for her own future burial and memorialization.[2]

Other women I interviewed spoke of myriad problems resulting from giving up their family name and taking their husband's surname in marriage. Of the 29 in-depth interviews I conducted with non-experts, 11 of the 24 women (there were three men and two transgender) had changed their name at least once. I will address the experiences of the 13 women who never changed their name in Chapter 3 when we consider common-law marriage as a strategy for keeping maiden names. Article 750 of the Civil Code requires that married couples share a common surname. This Article is at the heart of lawsuits over discrimination against women and debates about the meaning of marriage in Japan today since close to 97 percent of all marriages are registered under the male's surname.[3] Of the 11 women who changed their name in marriage, one is a foreigner who married a Japanese man and the other ten are Japanese citizens.

The history of family names

The historical development of family names and family identity in Japan sheds light on the surname debate in contemporary Japanese society. In particular, the linkage between geographical place, social status, group identity, and surname established in ancient times among elites and through the medieval period among samurai shows the tight association of people with their geographical, social, and kin-group identity, an identity that congeals in family name in the modern *koseki.*

Both male and female interviewees in my study and survey respondents showed a strong sense of belonging to their natal families and a strong sense of identification with their natal names. Correspondingly, female interviewees showed a strong sense of dislocation when required to relinquish their maiden names and assume their husbands' names. Reporting the dilemma of name change in marriage, one respondent in her mid-forties wrote, "I felt like the person I had been up until then was completely erased" (Survey 128). More than half of the approximately 75 survey respondents who had changed their name in marriage said they had suffered from a sense of loss, erasure, invisibility, or loneliness as they navigated their lives under a new identity.

Through interview and survey material I examined emic understandings of name and identity during ethnographic research in 2012–2013 to unravel the implications of both male-named family units and name-changing on women's lives. Material from interviews with married women who struggle to find an identity once they have left their natal *koseki* as well as stories from divorced women who have changed their names more than once, first in marriage and then divorce, expose the incommensurability between women's name experiences and men's. These stories reveal assumptions about women's names as changeable and women's identities as derivative of relationships within patriarchal frameworks that ultimately produce gender imbalances and fundamental aspects of gender in Japan. How are the women's experiences different from the experiences of the men they marry and divorce? How does name instability shape the experiences of single mothers or divorced women? Given women's central role in raising children – whether those women are married, unwed, or divorced – how does name-changing affect their positions as mothers in the world of their children? But first let's look back to the development of surname use in Japan and consider what names have meant historically.

Naming the Gods, naming the realm: the power of naming

According to Herbert Plutschow (1995), in ancient times, names were understood to be ontically linked to their referents. To speak a proper name was to conjure the thing or person named. Because of this, names, like the revered entities to which they applied, were treated with respect, power, and, sometimes, with fear. From the earliest records, deities of the Japanese islands were given names. The eighth-century *Kojiki*, creation story of the Japanese archipelago written entirely in Chinese characters by the Yamato elite, recounts the naming of deities who in turn name the other deities and the islands of Japan – creating and naming the islands simultaneously: "[T]hus the last pair of the heavenly eight deities, Izanagi and Izanami, by name, created the following 'islands': Awaji (Ogestushime), Sado (Takeyoriwake), Oki (Amenooshikorowake), Tsukushi (Shiraiwake)" (Plustchow 1995: 18). In the *Kojiki*, naming the islands brought them into physical form. The divine deity's creative power was manifest through the act of naming.

Early influence from the Asian mainland provided Japan's elite an orthography, a religious worldview, art, sculpture, and other tools that enabled Japanese society to develop rapidly. One of the many important influences brought from T'ang China in the seventh century was the legal system, known as the *ritsuryō*. This legal system provided bureaucratic structure and legal codes to facilitate a system of governance that depended on social and political hierarchy based on the Chinese model. The early government kept records of aristocrats, warriors, and others appointed to government positions. According to Plutschow, this meant that "we know much more about the names of the upper than about the lower classes" during this period. (Plutschow 1995: 159). Peasant names were not recorded with any regularity and peasant identities became closely linked to the fields the peasants worked through time. As part of a taxation system, revamped in the tenth century, "the government decided to levy taxes on arable land rather than on people and let these lands be subsumed into often privately owned manors whose administrators were responsible for taxation" (Plutschow 1995). The lands, known under names called *myō*, were also used to identify the owners or administrators who collected the taxes. Thus, these *myō* identified land and people working the land as, somehow, one and the same. As Plutschow argues, "the association of peasant with his land was so close that one cannot separate field name and personal name, *myō* constituting a kind of metonymical identity of both field and cultivator" (Plutschow 1995: 161). The *myō* tended to be auspicious names that included, for example, the character for rich, *tomi* as in the surnames Hisatomi and Shigetomi (Plutschow 1995: 161). Those who worked the named land could use the name of that land as long as taxes were paid on the land. These early practices reveal the strong linkage between place and family name, even though that linkage was not yet codified in a register of any kind. The term for the named fields, *myō* 名 is the first character in one of the Japanese terms for surname, *myōji* 名字. The character 名 can be pronounced *na* as in *namae*, 名前 referring to first name, *myō*, 名, referring to last name, or *mei* as part of the compound *seimei* 姓名 or full name. The Japanese character derived from Chinese suggests the connection between geographical location, in the form of a named field and name of the peasants working the field.

Although these *myō* were not recorded, they came with some status and continuity, and in some cases, it was possible upon death for a man's land and the name of the land to be inherited by his son (Plutschow 1995). Although peasants did not officially have surnames, many historical examples reveal commoners using a surname. In fact, "[i]n 1483, the Tō-ji temple (Kyoto) complained that too many peasants had assumed surnames without permission," suggesting a trend that threatened social stability. Just prior to the Tokugawa period in the late sixteenth century, as part of his campaign to establish strict segregation between samurai and commoners, the powerful *daimyō*, Hideyoshi Toyotomi, declared a prohibition on the use of surnames for all but the samurai in 1590, "[h]owever correct one's ancestral history may be, and however rich one may be, people who fell to the status of ordinary people may under no circumstances bear names or swords" ([*Chihō Bonrei Roku*], in Plutschow 1995: 169). This prohibition on surname use by commoners continued in the Tokugawa period, during which time, concern persisted that surname use by commoners would threaten samurai and shogunal hegemony. In another sign of the significance of names and naming regimes, the people with the most power in society, except the Shogun during the Tokugawa period, were referred to as *daimyo*, a term written with the characters for big name, 大名.

Although identity during Tokugawa may have been closely linked to land and social class category based on occupational status – as samurai, farmer, artisan, merchant, or outcaste – by 1871 social class distinctions were outlawed and the new Meiji government sought a new way to "grasp" the population. That "grasp" signaled the beginning of the *koseki*.[4] The *koseki* system became the new device for classifying the population within a modern society.

History of registries

Although population registration in Japan dates back to the Asuka period (592–710) the current *koseki* in use is a modified version of the *koseki* implemented in the Koseki Law of 1871 in the very early years of the Meiji Restoration.[5] In Meiji we see a mix of new western ideas mingling with Tokugawa practices. Because names had long been a sign of power or relationship to power, it made sense that the new government would now take over the role of *authorizing names*. Thus, within three years of the establishment of the Meiji government, the Koseki Law was implemented (passed in 1871 and put in force in 1872) and the new government required all commoners to take surnames in 1875.

According to Mori, many characteristics of the early Meiji *koseki* introduced through the rapidly passed Koseki Law should be understood as a response to the failure and inadequacies of earlier registration systems. During the Tokugawa period, population and family documents varied by area. In fact, two different registration systems existed at the same time. The *shūmon aratame* register was carried out at Buddhist temples beginning in the seventeenth century to squelch Christianity through forced registered affiliation with Buddhist temples (Cornell and Hayami 1986). These registers recorded household data by occupationally based social class and place of residence. They included details of deaths, births, marriages, and other vital family relationships and events. At the same time, a population census called the *ninbetsu aratame* was carried out periodically under the supervision of the samurai and higher authorities. This census tracked households by occupational status, as nobility, samurai, peasant, townspeople, and outcastes (Mori 2014: 63). During the Tokugawa period, the *ninbetsuchō* and *shūmon aratame* "were gradually consolidated and came to be referred to as *shūmon ninbetsu aratamechō*" (Mori 2014). But the consolidation left many gaps in information, in part because the registers could not easily account for the movement of people from one area of the country to another and did not encompass all events that occurred between census takings. This was further complicated by the absence of surnames for many households. The *shūmon aratame* in particular were organized by geographical location surrounding a temple. As such, a group of people were documented in relation to their dwelling on the land. The modern *koseki* is not very different. Since the development of its modern form, established in early the Meiji period, the *koseki* has been located in local government offices under the auspices of the national government, by address and surname of the household head. In this sense, the local government office has replaced the temple as the locus of information and governmental surveillance and power.

In the transition to Meiji when the system of occupation-based social identity was being dismantled, family name, and in particular a stable household name, began to take its place as a form of social location. By the time the Meiji government made a decree requiring all commoners to take a surname in 1875, many had already done so. To meet the surname requirement people used a range of strategies to choose a name for their families. Most commonly, they registered the name they informally or privately used, which was often a field name or distinctive natural feature, such as a nearby mountain or a unique or particularly large tree (Plutschow 1995). According to the social historian, Yanagita Kunio, commoners experienced anxiety in early Meiji about choosing names because they worried about offending superiors within their communities. This fear was a natural outgrowth of prior taboos on using a name that shared a component (one of the characters in what was usually a two-character compound) with the name of a lord or superior. Commoners also worried that the law would change again in the future and that their names might be taken away. These concerns drove some villagers to their local priests for guidance in choosing names that were propitious and safe in the early Meiji period (Plutschow 1995).

In Meiji, the site of registration of families moved from Buddhist temples to local government offices and responsibility for completed registrations moved from local samurai or land managers during Tokugawa, to government clerks and newly authorized male heads of household, *koshu*. The term *koshu* (rendered as 戸主) is an amalgam of *ko* of *koseki*, the counter for houses or families, and *shu* or master (as in *shujin*, a somewhat outdated term for husband), thus, the term means master of the house or head of the household. This amalgam, which is often translated as household head, reveals the hierarchical structure of the family and authoritative nature of the *koshu*. In the new Meiji state, the *koshu* was fashioned as not only head of the family, but as a go-between, linking the family unit to the newly formed national government. The empowered *koshu* served the new nation and the growing imperial state which extended governmental rule to the family unit through the *koshu*'s role and status. This linkage ultimately raised the profile of the household head, while formalizing the hierarchy within the family under his supervision. The household head's reputation depended on his family's reputation. Doing his job well meant keeping his family under control, and that included keeping the *koseki* accurate and up to date. Thus, the content of the *koseki*, like the content of the household, was the responsibility of the head of household. As such, marriages of children or dependent and in-house siblings were the head of household's responsibility and accurately reporting any changes to the household structure to the local government authorities was not only required by law in the 1871 Koseki Law, but was also understood as family and national duty. Government-authorized responsibility to keep the *koseki* accurate and updated served to strengthen household hierarchy and patriarchy. This burgeoning modern structure marks the significant interpenetration of the public and the private realms in early Meiji Japan (Mackie 2014).

Linking *koseki* to *ie* and *ie* to *koseki*

The *ie seido*, or household system, has received much attention in Japanese studies. The *ie* refers to a particular type of household or family structure that became fixed in law in the early Meiji period. We can define the *ie* simply as a multiple-generational family ideal that guarantees continuity and inheritance, generally through the eldest son, but in some cases through the youngest or eldest daughter. Much debate, however, surrounds the sources of the modern form of *ie*, and whether it is rooted in feudal Tokugawa family structure or an entirely modern structure more concerned with nineteenth-century western notions of family.[6] Discourses on the *ie* reflect the potency of this model family within the field of Japanese studies. Understanding the essential components of the *ie* has been interpreted as essential for understanding contemporary society and gender problems today.

While the *ie* took shape during Meiji, scholars of the *koseki* system point out that the *koseki* system, including the laws and the document and the way the document was processed, became constitutive of the *ie* family structure, "the '*ie*' was first defined by the *koseki* as a family group. It was only the *koseki* that made the *ie* discernible to the eyes of the nation" (Ninomiya 2014: 175). While it has often been said that the *koseki* maps the *ie*, Ninomiya argues that the Koseki Law and, thus, *koseki* system in fact specified what later became known as the *ie seido*. "In this way, the *koseki* was not a mere register of family relations, but rather it exemplified and actualized the model that constituted the foundations of the family in the Meiji Civil Code" (Ninomiya 2014: 175). Feminist sociologist, Ueno Chizuko, has argued that the Meiji era *ie seido* imposed new expectations on succession and family structure that expanded the hierarchical samurai model to all other members of society. So while hierarchy and patriarchy certainly existed before the Meiji period, the Meiji form of the modern *koseki* expanded upon that hierarchy and patriarchy, essentially requiring it of all families.

The new family form, which was codified in the Koseki Law of 1871 and in the *koseki* registration process, resulted in numerous negative consequences for women, regardless of socio-economic or occupational class status. Prior to Meiji, strategies for succession varied across the country (Toshitani 1994). While patrilineality was common, eldest son inheritance or succession was most common in the samurai class and was not dominant in all classes and regions. In fact, eldest or youngest daughter inheritance was not uncommon, particularly in rural regions of the country (Toshitani 1994). It was a minority of Japanese, the elite samurai, less than 10 percent of all families, which depended on patrilineal descent, arranged marriages, and primogeniture to increase family prominence and success. This samurai-like *ie* (household) form was imposed on all citizens through the 1871 Koseki Law, by subsuming all family members under one head of household, the *koshu*, almost always a male, mimicking the samurai norm (Toshitani 1994).

In the Meiji period, a woman's marriage was often referred to by the expression *koseki ni hairu*, "entering a *koseki*," or *oyome ni iku*, "going as a daughter-in-law." Both of these oft-used phrases reference a movement from one house to another and one *koseki* to another. Entering the *koseki* or going as a daughter-in-law means leaving a natal home and family and becoming part of another family/*koseki*. In the new consolidation of family and law in the Meiji period, this entering into a household required that a woman be struck from her natal *koseki* (marked by a large X called a *batsu* that is also used for the divorced and deceased) and entered into the husband's family's *koseki* as wife of son, with only her first name listed. In a column next to her given name, the name and place of residence of her parents was written. From the time of entering her husband's family's *koseki* she would assume his surname, though in many cases, that surname would never be used when referring to her, since her appellation would often be "wife," or simply the term, Mrs. (*okusan*) the household (her husband's) surname. She was understood as essentially property of the husband's family (not just the husband) (Smith 1987). Meiji-era *koseki* ranked all members in order of their entry into the family. An eldest son was already listed in his parent's or grandparent's *koseki*. His marrying-in wife was listed at the end of the *koseki* as the most recent arrival. Children born to her would be listed in order of birth after her. In this way, all family members were listed in the *koseki* in hierarchical order and this ranking on the *koseki* reflected material conditions of hierarchy in the family. Usually the wife of an eldest son would serve the entire household as daughter-in-law/wife and provide a wide range of services to her mother-in-law and father-in-law as well as siblings of her husband and any other children (siblings of the husband) in the home. Marriage was not primarily a relationship between a man and a woman, but between a son's household and a marrying-in wife/daughter-in-law.

The rules governing the content of the *koseki* reinforced the importance of names and naming. Many articles of the 1871 law as well as the revised 1947 Koseki Law, which currently stands, stipulate *koseki* changes and *koseki* rules by way of the status of names. For example, in Article 9, "A family-register shall be indicated by the full name and registered locality of the person who appears first in the family-register. The same shall also apply after such a person has been struck off a family-register" (EHS LW Bulletin Series Vol. II) and Article 18, "A child who assumes the surname of its father and mother shall be entered in the family-register of the father and mother" (EHS LW Bulletin Series Vol. II). Quite simply, the law is centered on the premise that every household assumes one surname. In Meiji, the head of household was expected to be a male, fitting with the model of the samurai, patriarchal household, which became the normative standard in the Meiji period in part because it had been the norm for most of the young men forming the new government who were former samurai (Ueno 1987).

Persistence of the *ie seido* logic in the modern *koseki* document

Meiji-era *koseki* were of course handwritten vertically from right to left in the Japanese style. Today's *koseki* are printed in western style, typed horizontally on a form that is read from left to right. Figure 1.1 shows a sample of a typical Meiji period *koseki*, where the defining feature is the address and full name of the household head written vertically on the far right side of the document (the position of the beginning of the *koseki*). Figure 1.2 shows a sample of a recent, horizontal and computerized *koseki* document.[7]

Although these two documents are strikingly different in form and appearance because of the lapse of over 130 years, they are also surprisingly similar. Feminist critics, such as Sakakibara Fujiko and Tanaka Sumiko, argue that the *koseki* document used today carries the remnants of the Meiji-era *koseki* and the Meiji-era *ie seido*, household system that it codified (Sakakibara 1992; Tanaka 2004a; Ueno 2009). They contend that we have a system of registration made for nineteenth-century society with assumptions about gender and the family from the nineteenth century still in operation in the twenty-first century.

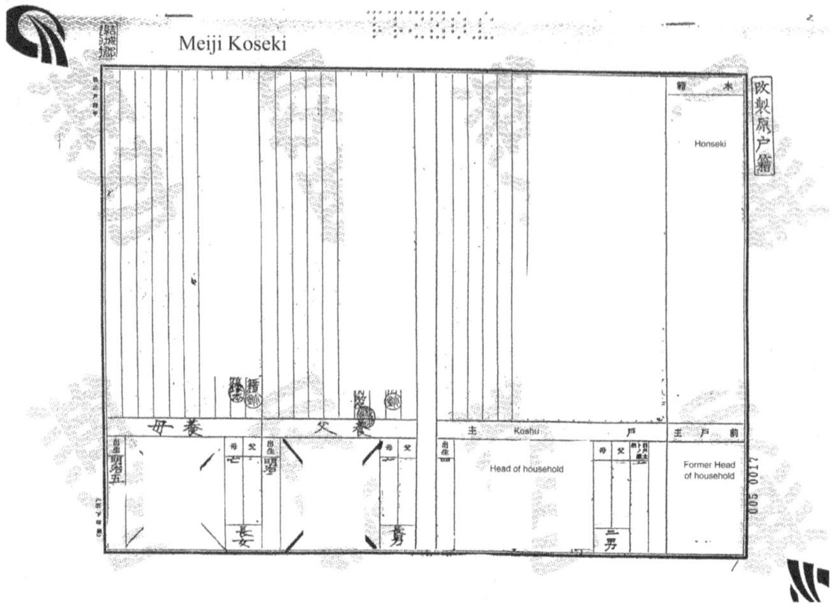

Figure 1.1 Meiji period *koseki*.

There are five main features of the Meiji-era *koseki* that have remained in the present-day *koseki* document. These include:

1 indexing the household by location using a permanent address called *honseki*;
2 listing the household surname in the first column of the document just beyond the address as an identifier of the household;
3 listing each family member, whether spouse or child, in an individual section by first name only after the head of household in order of entrance into the household through marriage, birth, or adoption;
4 specifying each person listed on the *koseki* by his or her relationship to the head of household, as spouse, eldest daughter, eldest son, and so forth;
5 assuming that the first name on the *koseki* will be the male household head (although this is not required).

These continuities in structure and information specifications help transmit understandings of gender and family across several centuries.

The primary differences, insofar as gender is concerned, between the Meiji *koseki* and the *koseki* in use today include:

1 the change of the term for the head of household from *koshu* (house master or head of household, 戸主) to *hittōsha* (first registrant, 筆頭者), where the

Parental Koseki

全部事項証明

本　籍 Honseki	群馬県前橋市千代田町一丁目２７番地
氏　名 Name	丙原　隆二　**Heihara Ryuji**
戸籍事項 戸籍編製	省略
戸籍に記録されている者	【名】隆二　**Ryuji** 【生年月日】昭和５１年１２月３日　【配偶者区分】夫 【父】丙原信吉 【母】丙原夏子 【続柄】二男
身分事項 出　生	省略
婚　姻	省略
戸籍に記録されている者	【名】竹子　**Takeko** 【生年月日】昭和５４年５月１２日　【配偶者区分】妻 【父】甲原忠太郎 【母】甲原杉子 【続柄】長女
身分事項 出　生	省略
婚　姻	省略
戸籍に記録されている者	【名】信子　**Nobuko** 【生年月日】平成２４年１月１０日 【父】丙原隆二 【母】丙原竹子 【続柄】長女
身分事項 出　生	【出生日】平成２４年１月１０日 【出生地】仙台市青葉区 【届出日】平成２４年１月１６日 【届出人】父 【送付を受けた日】平成２４年１月１８日 【受理者】仙台市青葉区長
	以下余白

発行番号 000001

（例二）父母の戸籍

－ 34 －

Figure 1.2 Marital *koseki.*

term *koshu* is laden with patriarchal significance while *hittōsha* has a legalistic and gender neutral connotation; and

2 the change in 1947 of the *koseki* from a multiple generational record during the Meiji period to a two-generational record reflected in Article 16 of the *Koseki* Law, which states: "If a notification of marriage is given, a new family register shall be made up for a husband and wife" (2008).[8]

Abolishing the hierarchical, patrilineal family structure of the *ie seido* was a major goal of the Allied Occupation of 1945–1952. This goal sprang from a general belief that more democratic family relationships would help produce a more democratic society (Satō 1996; Dower 1999; Gordon 1997). Satō Bunmei points out that the Constitution committee of the Occupation attempted to abolish the *ie seido* by granting women equal rights in society in Article 14 and equal rights between husband and wife in Article 24 of the 1947 Constitution. However, after much behind-the-scenes committee debate by Japanese government representatives and Occupation members about whether or not to abolish the *koseki*, although the relatively progressive Articles to the Constitution were accepted, the *koseki* was left in place, leading to contradictions between the Constitution and the Koseki Law that continue today. These historical contradictions born during the Occupation persist today as equality is guaranteed in the Constitution while hierarchy is preserved in the *koseki* system.

Patriarchal remnants in the *koseki* structure today

Although simple in style, the *koseki* holds a wealth of information. This model *koseki* is taken from the *Koseki jitsumuno tehiki* 2011, used by city workers in order to comply with the Koseki Law and the Civil Code.[9] In Figure 1.3 a *koseki* for a family with three members is represented. The sample is titled "Parental *koseki*" and reads vertically like a Meiji-era *koseki* from right to left. The first column is titled *honseki*, 籍本 (legal domicile) read right to left, with the address written vertically below.[10] Under the address, the title *shimei*, 名氏 (full name) is listed horizontally and below that the *hittōsha* (first registrant or householder) is named.[11] The column next to the *honseki* states that other matters have been abridged on this version of the *koseki*. In other words, more information is available on a fuller version of the *koseki* which might include other marriages, divorces, adoptions, and so forth. The next column is information about the first person named on the *koseki*, the *hittōsha*. Birth and marital information about this person has been abridged as stated at the top of the column. Below that, in the same column, the first name of the *hittōsha*, Ryūji (隆二) is given in the center of the section labeled husband 夫 (*otto*). In this part of the *koseki* the name of Ryūji's father is provided in full, 内原信吉 and mother's first name, 夏子 Natsuko. Below the parent's names is the *tsuzukigara* for Ryūji, 男二. *Tsuzukigara* refers to the column on the *koseki* where gender and birth order are specified. We can see that Ryūji is the second son, *jinan*. To the left of his name

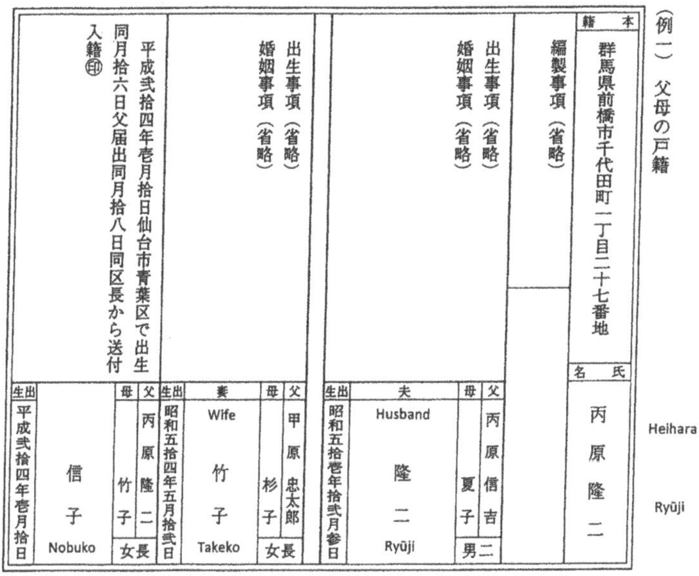
（例一）父母の戸籍

本籍 群馬県前橋市千代田町一丁目二十七番地

氏名 丙原隆二 Heihara Ryūji

編製事項（省略）

出生事項（省略）
婚姻事項（省略）

父 丙原信吉 母 夏子 二男
夫 隆二 Husband Ryūji
生出 昭和五拾壱年拾弐月参日

出生事項（省略）
婚姻事項（省略）

父 甲原忠太郎 母 杉子 長女
妻 竹子 Wife Takeko
生出 昭和五拾四年五月拾弐日

平成弐拾四年壱月拾日仙台市青葉区で出生
同月拾六日父届出同月拾八日同区長から送付
入籍㊞

父 丙原隆二 母 竹子 長女
信子 Nobuko
生出 平成弐拾四年壱月拾日

Figure 1.3 Parental *koseki.*

is his birth date. This column of the *koseki* provides all the information about this family member, Ryūji. In the next column, in similar fashion to the husband's column, is the information about his wife, Takeko. Just as the details related to Ryūji's birth and marriage are abridged in this document, so are hers. At the bottom of her column we see the term wife 妻 *tsuma* and her first name, Takeko 竹子 below that. To the right of her name is her father's name given in full and her mother's first name. Her *tsuzukigara* is given below her parent's names as eldest daughter 長女 *chōjo*. On the left side of this section is her birth date. The next person listed on the *koseki* is Takeko and Ryūji's child, named Nobuko 信子. Because this is the new information on this particular sample *koseki* the full dates of the birth and the notification by her father and mother are written above her name.

As we can see in the sample *koseki*, the *tsuzukigara* is provided for everyone named on the *koseki*, including the head of household, the wife, and the child. The terms for these gender and birth order family positions are rendered in English as eldest son (*chōnan*), eldest daughter (*chōjo*), second son (*jinan*), second daughter (*jijo*), and so forth. The word *tsuzukigara* may be translated as "family position" or "order of succession," but that does not quite capture its significance in rendering family relations historically and in the present. The *tsuzukigara* column represents the relationship between the head of household and the individual named, and as such, it is a remnant of the important role of succession in the Meiji-era *ie seido* where primogeniture, eldest son inheritance, was the legal norm. This form of succession where one child would take over the headship of his parents' home and work (whether farm, shop or other type of business) while all other children were expected to marry out and leave shaped many facets of understandings of family. The fact that the *tsuzukigara* column continues to be a prominent feature of all *koseki* reveals the ongoing significance of birth order and gender in understandings of identity and family. This feature of the *koseki* and the language used to specify birth order will be discussed further in Chapter 3 because it plays a critical role in several lawsuits regarding the legitimacy status of children born to couples in common-law marriages.

One of the central aspects of the Meiji-era *ie seido* and *koseki* was that it linked multiple generations through a single last name that stood for the household or family and thus served as a record of family history. When the 1947 Koseki Law specified an end to this structure, it provided a version of family reckoning that appears to specify only the nuclear family in two generations. However, because parental names, birth order of children, and place of birth are all provided for each person listed in every *koseki* it is very easy to track multiple generations of both the husband's and the wife's family on the postwar two-generation *koseki.*

New privacy laws enacted in the 1990s have reduced the previously easy access to non-kin *koseki* documents. But until the privacy laws were enacted, knowing that an employer, potential in-law, or business partner could easily access one's *koseki* played a powerful role in convincing people that at all costs they must keep the *koseki* unsullied, *kirei.* A stain, such as divorce, non-marriage, death of a young person, and so forth, on the *koseki* was like a stain on the family name and this sensibility continues today even though access to another person's *koseki* has become more restricted. One of the privacy stipulations is defined in Article 10 of the Koseki Law, which limits a non-family member's right to see a *koseki,* stating that only a person appearing on a *koseki* or the spouse of such a person or lineal ascendant or descendant may request a copy or an abstract of a family register. Beyond that, other parties may request access to a non-familial *koseki* only under very circumscribed conditions. The control of all *koseki* is left up to the mayor of the city, town, or village, giving the representative of the state great authority in the dissemination of personal information on the *koseki.* Prior to the addition of Article 10, and a heightened sense of privacy in society generally, citizens were made to understand that their family documents were records that could be obtained by any interested party. The sense that others were watching in small-town Japan was actually backed up through this administrative system that would allow a detective, a future in-law, future employer, or other interested party to see the family information. Daughters' marriages figured centrally in the information on a *koseki.* A daughter who did not marry in a timely fashion was suspect, as was a wife who did not bear children. In this way, the *koseki* had a normalizing effect on many stages of women's lives.

Although changes to the Koseki Law and the Civil Code during the Allied Occupation of Japan purportedly resulted in the abolition of the *ie seido* family structure, it is easy to see *ie seido* structure extant in the *koseki* above. *Koseki* are still indexed by address and surname. Address has been a key part of ordering and understanding family. Even families who have lived in the Tokyo area for many generations may register their *koseki* in the village where the ancestors lived and perhaps where property or a house is still owned. But even when the relationship to the land has long since dissolved, people may keep their registers in their ancestral hometown, because that is where the deceased family members are often enshrined in rural Buddhist temples where family members like Nakayama return for seasonal memorial services. It is not uncommon to feel that the *koseki* should be kept where the ancestors are.

This connection between place and name and family is solidified through the *koseki* document, but also through funerary and memorial practices closely associated with family and *koseki*. Children are obligated to care for the spirits of their ancestors through rituals at Buddhist temples and in their homes with household altars where some of the ashes may be kept. These memorialization tasks usually fall to women. Traditionally, every household had a daughter-in-law to take on this ritual work of ancestor care, but increasingly because of the shrinking family size in the last several generations, daughters, whether married out of their natal home or not, become responsible for their own parents' memorializations. Most elderly are cremated and memorialized in the Buddhist temple close to their home where their own parents or in-laws are memorialized. In this way, the funeral practices shape understandings of home and family and may be reflected on the *koseki* in the *honseki* address (Smith 1974).

The impact of name changing on a woman's life

When a woman changes her name upon marriage in Japan today and moves into a *koseki* registered under her husband's name, even though it is a new marital *koseki*, the *honseki* address and the *tsuzukigara* column linking her husband to his parents and their surname confirms the earlier Meiji-era understanding that she is an outsider coming in to her husband's family line with the obligations of an outsider "daughter-in-law." Tanaka Sumiko, an activist who has never changed her name and whose lawsuit over discrimination against children born out of wedlock in the *tsuzukigara* column of the *koseki*, argues that understandings and attitudes about marriage and the role of daughters-in-law are still embedded in the *koseki* structure reflected in the terminology and structure of the *koseki*.[12] In other words, when a Japanese wife changes her name to her husband's, she is doing more than changing her name. She is participating in a system that subordinates her identity in her husband's family structure. This subordination occurs through the requirements of *koseki* registration, specified in law, which then become the basis for ideological understandings of marriage, family and identity. This system of family reckoning has not substantively changed since the Meiji era and is codified in the *koseki*.[13]

Mikiko Ishihara, another key informant and friend, provides another view of the relationship between *koseki* and marriage. She first married when she was in her mid-twenties. She was eager to "retire" from her job, and happy to take her husband's name, she told me. At the time, in the 1980s, a husband with a solid, white-collar job and a nice apartment in Tokyo was a good catch and she, at least initially, welcomed married life. However, after the birth of two children and because of a growing sense of dissatisfaction with her marriage she filed for divorce and became a single mother before the children entered elementary school. No sooner had she been released from an unhappy marriage than she faced the problem of her name. The children had her husband's name and she did not really want to take back her maiden name and have a different name than her children. Going back to a maiden name was not particularly accepted at the time, she told me. People saw daughters returning to their parents' *koseki* or their maiden names as failures. In addition, she liked her marital name and wanted to have the same name as her children. However, at the time of her divorce, women were expected to resume use of their maiden name, literally returning to their father's *koseki* in a process that reflected an understanding that women must be attached to a male-headed household in the role of daughter, wife, or elderly grandmother obedient to father, husband, or son, respectively. Mikiko had to file a petition to get permission to continue using the name Ishihara after the divorce. In fact, she had to change her name and re-register as Ishihara using a name-change process so that she would no longer be connected to her ex-husband's name Ishihara. Ishihara is a common enough name and the judge granted her request and the *koseki* which she now heads with the name Ishihara Mikiko shows no connection to her former husband, except that her children's names on her *koseki* show her ex-husband's name under the father column of their sections of the *koseki*. Mikiko told me that it was inconvenient to change her name and file the petition with the court. She had to pay court fees and there was a waiting period when she was not sure which name she would have. In the end, she was glad to have kept the name Ishihara because it allowed her to have the same name as her children after the divorce without disrupting their lives at school or in other settings where a name change would have drawn unwanted attention to the divorce. Furthermore, using the name Ishihara protected her own parents who would not be required to welcome their daughter and her children into their own *koseki*.[14]

Several of the divorced women I interviewed who changed their names upon marriage were not as sanguine about using their husband's name in the first place and filed for paper divorces to regain the use of their maiden names in the early years of their marriages. Sumitani Yukiko, a bright young woman in her thirties, told me that she changed her name back to her maiden name several years after she was married by getting a *peipā rikon*, a paper divorce. A paper divorce is obtained by going to the local city office where the *koseki* is registered and submitting a Divorce Notification form. The divorce goes into effect immediately and a new *koseki* is issued for the wife who is crossed off the marital *koseki* headed by the husband. The husband does not have to start a new *koseki* because his name does not change. But divorce allows the wife to resume her maiden name, which is the point of the process. Sumitani legally changed her name through this process and became the *hittōsha* of a new individual *koseki*. On the *koseki* headed by her husband there is a large X mark through his wife's name and in the information column above her name, the date of divorce is given as is standard for divorce, because a man doesn't usually change his name upon divorce, only a wife leaves a *koseki*, resuming a maiden name. Thus, Sumitani's husband appears divorced and could legally remarry. Because the couple had no children, her husband was then alone in his *koseki* and Sumitani was the sole individual in her newly created *koseki*. The practice is called paper divorce because couples who employ this practice understand themselves as still married even though their legal status is divorced. Such paper divorces are one of the clever workarounds increasingly popular with women who are trying to keep their maiden names in marriage. Sumitani told me that her husband understood her reason for wanting to regain her maiden name and said he did not oppose the paper divorce. However, Sumitani's boss was not as understanding about her change in status and as soon as she notified him of her name change, problems started at work. She said "the climate at work changed immediately after I told my boss that I had changed my name." She told me that she was quite certain that the cool treatment she received was due to her name change. I asked her why changing her name would cause such a problem at work and she responded, "Japanese companies like *sunao na hito*" (gentle, compliant people who will not rock the boat or cause problems). "I became perceived as someone who was egotistical and who would challenge the social norms." About five years after their paper divorce Sumitani and her husband decided to actually part ways, in other words, really divorce. She said their relationship had become more like brother and sister and she was frustrated and wanted to find a new relationship. She also admitted that since they were already divorced on paper it was easier to separate and truly break up.[15]

Several other women I interviewed had chosen paper divorce in order to legally assume their maiden names. In some cases this was a perfect solution to the name problem. One woman with children did not mind the fact that she had a paper divorce because her children had already moved out of the marital *koseki* for their own marriages before she filed the divorce notification to reclaim her name. Paper divorce allowed her to legally use her maiden name. Because of her age and the long-term stability of her marriage, she said that she did not feel insecure about the lack of proof of her marriage.

One of the most fascinating stories I heard about the inconveniences and disruptions of marital name-changing came from Hiroyo Maruyama, whom I met at the Single Mother's Christmas Party in December 2012. Maruyama-san had been a single mother until recently. Originally married to a foreigner and now married to a Japanese man, her teenage daughter had dual citizenship. Maruyama-san told me about her teenage daughter's plan to study abroad in the United States and hoped to further discuss the issue with me. She was concerned about the academic consequences of studying in the United States and wanted my opinion about a particular school in the Midwest, which her daughter planned to attend. That conversation provided a context for ongoing conversations prior to the interview in January. Maruyama-san had recently remarried and for the first time in her life changed her surname (although she continued to use her maiden name in her dealings with me and other acquaintances). Her first marriage, over 15 years earlier to a foreigner, meant that she did not have to change her name for that marriage. Foreign spouses figure only minorly on the *koseki* and can never be the *hittōsha* of the family. The single surname requirement and other associated effects of heterosexual marriage in Japan, in fact, only apply when both parties are Japanese citizens. Because foreigners do not have the citizen's household register they do not significantly alter a household register when they marry a Japanese. Through the Meiji Koseki Law, the *koseki* demarcated both family membership and national membership. This dual purpose has resulted in strange and contradictory effects. As Mori Kenji notes, "the use of the *koseki* as documentation of nationality in the modern period constitutes an important new purpose. Many paradoxes and contradictions have arisen from nationality being documented through the *ko* [family or household] unit" (2014: 61). Satō Bunmei (1996) Karl Jakob Krogness (2011), and David Chapman (2014) all elaborate on these paradoxes and contradictions.[16] So ironically, the document that signals and reports family relationships (husband, wife, and children) does not perform that essential function when the husband and wife are not both Japanese citizens. This contradiction between *koseki* as record of nationality and record of family was born in the Meiji Koseki Law and has not yet been eradicated.

The contradictions of family and nationality tracking in the *koseki* hit home in Maruyama-san's story. During her first marriage, she clearly enjoyed the privilege, one familiar to Japanese men, but not women, of keeping her own name. She conveyed the significance of the paradox of gender and citizenship that shaped her first marital *koseki* when she said, "*toppu datta*" ("I was on the top of the *koseki*"). Because she was the "toppu" of her first marital *koseki*, her daughter's name is also Maruyama, in keeping with the single-surname rule. Her first husband did not take her name because he, in fact, had no standing as a family member as far as the Koseki Law was concerned. His name was listed in a marginal position on the *koseki* document and the date of marriage given, but significantly, his status was not like a Japanese citizen and thus his name did not have to change.[17] The column after the Japanese wife's entry is blank and the information about the marriage to the foreigner, Wayne Van den Bosch, is listed above her name, rather than in the column next to her as the husband.

When she married her current husband, Maruyama-san told me, "*koseki ni haita*" ("I entered or became part of his *koseki*"). This meant that for the first time in her life, in her late thirties, with a high-school-aged daughter of the same surname, she would have to change her name. I did not have to ask about whether or not she and her prospective husband had discussed the possibility of separate surnames or of her husband taking her surname, because she told me immediately that his family business made it impossible for him to change his name.

But the description Maruyama offered of the process of changing her legal name on her credit cards, employment records, insurance cards, and daughter's school records was hilarious. She was terribly put out by the inconvenience, like someone who assumed her name would always be the same, never knowing anything but the normalcy of keeping her name. More than anything, she sounded like the Japanese men I had interviewed who took their names and the continuity through time of their surnames for granted. She had been privileged in her adult life never knowing the frustration and inconveniences of changing her name that so many of my interviewees discussed. She had taken the ease of her identity and the authority of her name for granted. Suddenly, when confronted with a name change for her second marriage, she found it outrageous that she had to spend so much time in so many different bureaucratic settings, showing proof of her marriage and name change. "*Taihen deshita*" ("It was terrible!") "I had to go to so many different places and answer so many questions."

There are other important themes that emerge from the stories of women who changed their name in marriage. One problem mentioned by many of my survey respondents as well as in interviews was the problem of children's names after a divorce. Depending on the age of the children at the time of divorce and the relationship with the ex-husband (which tended to be rather negative), the disparity in names between mother and children was a big concern. A divorced woman may keep her married name or resume her previous name upon divorce. Either way, she must exit the marital *koseki* and create a new *koseki*. Some divorced women move their children into the new *koseki* with them, but not all, and custody is not determined by the *koseki* location. Since all members of a *koseki* must share the same name, if the children move to the mother's *koseki* designated by her maiden name, they all take her maiden name. Such a name change for children can be inconvenient or troubling, revealing the parents' divorce, which was seen as especially problematic for school-aged children by many of the survey respondents. It is possible to keep children in a father's *koseki* under his surname even as they move to a new living arrangement with their mother when she has full custody, but for a number of reasons this is not necessarily desirable. Japanese law does not currently allow joint custody of children after divorce which aggravates decisions about name and *koseki* location post-divorce.

While divorces can be complex and messy in any society, legal divorce in Japan can be accomplished very simply on paper when it is a consensual divorce by submitting a Divorce Notification to the local city authorities without a lawyer. In this sense the legal process is simple and immediate. Upon notification of divorce at the city office the *koseki* are changed and the new status (not married) is official. However, the implications of name change and registration of children can be quite complex.

The longer I study the impact of the *koseki* on divorce, the more I am inclined to believe that the *koseki* system itself is one of the structures in Japanese society that depresses the divorce rate, by making it very difficult to divorce because of the myriad name and status ramifications for children (not to mention for wives).

So while materially, the paper divorce is an easy process, on another level it is deeply complex. The curious phenomenon of *kateinai rikon* (in-house divorce) may also be explained in part by the inconvenient aspects of legal divorce in the *koseki*. Because an in-house divorce allows a couple to acknowledge or agree to separation (end of relationship) while keeping an intact legal marriage, they can end their intimate and personal relationship without showing evidence of divorce to the wider society, in other words, through divorce notification which would show on both parties' *koseki*. Generally, couples who have an in-house divorce stay together under one roof, but they are perhaps already estranged within that one home or essentially living separately because of employment transfer or some other thing that keeps them apart most of the time. There are in some, perhaps many cases, economic and social reasons to have an in-house divorce. But there is also a legal and *koseki* incentive, because divorce creates the problems of surname change and registration of children. Legal divorce also links back to parents and other relatives through *koseki* names and addresses. In this way, divorce is never just about a couple, but rather the reputations of both partners' families including ancestors and descendants. While paper divorce may be easy to explain, real divorce may be perceived as a sign of selfish individualism, where a person, or indeed, a couple puts themselves ahead of their families. The sense that all life decisions impact the larger family starts with the *koseki* where individuals and family share space in a network of documents linked through names, birth orders, and addresses. Divorce is one of those *koseki*-sullying events that many people want to avoid because it continues to be viewed negatively in society, even though this has changed greatly in recent decades.

In-house divorce may be an arrangement that is difficult for some to imagine, but it is a fairly popular strategy for separation/divorce in Japan. One woman I interviewed over 20 years ago talked about how happy she was to have that kind of relationship. She said that her husband was rarely home when she was and it allowed them to stay in their beautiful home.[18] Another key benefit for her was not having to report the divorce to family or friends. She appeared to be married and received social and financial benefits of marriage. Such marriages were common enough and considered quite normal at the time. She had no need to change her name, her *koseki*, or his official marital status even though their intimate and personal marital relationship had ended by mutual agreement. The inhibition regarding altering a *koseki* in divorce in particular must be understood in the context of family relationships and family status. Because *koseki* are understood to reflect families in more than one generation, the behavior of married or divorced individuals has consequences for the broader extended family. Just as the Meiji drafters of the Koseki Law had hoped, embedded in the structure of the *koseki* system were incentives to conform to societal norms like stable, long-term marriages (happy or not) and reproduction of a household heir (fertile or not). These documents and the events listed on them are understood to affect the reputations of all family members, as many of the people I interviewed often reminded me. Through my interviewee's ability to have an in-house divorce, her parents, children, and siblings and her husband's parents and siblings, not to mention other relatives, were spared the discomfort or social humiliation of divorce.

Conclusion

The significance of a surname varies across historical periods, across cultures, and within cultures. Many people are quite satisfied with patrilineal surname traditions and they have served many functions throughout history. In the United States, where it is possible for couples to maintain separate surnames in marriage (for women to keep their maiden names), most do not. But the difference between holding a legal right to maintain separate surnames and not holding such a right is profound. The most determined of the activists I know in Japan would never suggest that all women should keep their own surname in marriage. Rather, they argue that Japanese women should have the option to easily and legally maintain their natal surname throughout their entire lives, without imposing a change on their husbands. It is probably fair to say that most Japanese women would not keep their maiden names in marriage given the option. But for the women discussed in the next chapter who sued the government for damages due to its failure to change the law requiring one surname per family, women's names matter very much and the loss of identity and sense of self they describe has fueled their determination to gain the right to keep their natal names in marriage.

Notes

1 Thank you to Carolyn Sachs, my cousin and a rural sociologist, for suggesting that I emphasize the fact that there is widespread use of surnames in Japanese society today.
2 I do not know if the practice at her family temple in Shizuoka is particularly standard. It seems like a commonplace that wives are expected to be buried with their husband's family, and women as daughters and daughters-in-law do much of the memorial care for in-laws and ancestors in the marital line. Increasingly married daughters also provide ritual memorialization duties for their natal families and ancestor care for both their husband's family and their natal family for a variety of reasons including family feeling, changing attitudes about memorialization, increased overseas residence which leaves fewer people in Japan to care for the ancestors, and of course, the declining number of children in each family.
3 I actually interviewed one activist who had the good fortune to marry a man with the same last name. As such, she is the rare example of a woman whose maintenance of her own surname did not affect the surname of her husband. In other words, they both kept their surnames.
4 Kenji Mori emphasizes the need for the Meiji government to account for, or as he says, "grasp" those under its control for the purposes of conscription, taxation, and social control, through the family unit structure. "The Japanese *koseki* is an official document produced by the ruling authority to grasp those under its control (the governed or the *kokumin)* through the *ko* unit" (Mori 2014: 61).
5 The oldest surviving register, discovered in Fukuoka in 2012, is made of wood and contains names, information about military status, gender, and place of registration (*Yomiuri Shimbun*, June 14, 2012).
6 See for example, Hayakawa Noriyo (2014) and Ueno (2009).

7 The computerization of all *koseki* is still underway. Original *koseki* documents are being archived, but I was told by several sources that more than 90 percent of *koseki* are now computerized.

8 *The Family Registration Law of Japan*, EHS Law Bulletin Series, Vol. II, No. 2150.

9 I am indebted to Tanaka Sumiko who gave me her personal copy of the handbook, *Koseki jitsumuno tehiki* 2011, that she had acquired after years of employment at the Musashino City Hall. The book is compiled by a working group of lawyers for use by *koseki* workers in local government offices.

10 The *honseki* does not have to be the true current residence of the family, but the *koseki* will be kept and updated in the municipal office closest to the *honseki* address listed on the document. *Honseki, hon* = 本 *and seki* = 籍 and *shimei*, shi = 氏 and *mei* = 名 are written horizontally, right to left on the form.

11 There is much more to say about *honseki*, in part because it does not need to correspond to a real address and because of the linkage between land, geographical location and family identity. Ninomiya discusses some of the reasons people may move their *honseki* to addresses that are not their own, which would seem like a breach of the entire registration system, in order to hide certain personal information, such as, change of gender or divorce. See Ninomiya (2014) for an excellent discussion of this issue.

12 See White (2014) for a thorough discussion of the two lawsuits filed by Tanaka and her partner, Fukukita, against Musashino City. Chapter 3 will focus on the lawsuits and claims made against the government by not only these two plaintiffs but others who have assumed common-law marriages in order to avoid changing their names and the concomitant effects of a woman's "entering a *koseki*" in marriage.

13 Non-marriage or marrying a man who would take her surname would certainly change the situation of her social and familial status.

14 After divorce it would have been possible to re-enter her parents' *koseki* or to start a new *koseki* with her maiden name.

15 While *kami rikon* is often discussed by activists as a relatively easy fix for the surname problem, Sumitani's experience suggests that instability in both her work life and primary relationship resulted from changing her name and *koseki* status.

16 When I asked the *koseki* division supervisor about this apparent contradiction at my local city hall in Tokyo, I was told, "*koseki* are only for Japanese citizens" ("*nihonjin no tame desu kara*").

17 See Figure 5.2, a sample of material *koseki* where the husband is a foreigner.

18 She did not mention the impact a legal divorce would have had on her husband's career or their children's future marriages. Nor did she discuss a range of other critical consequences of legal divorce. However, clearly, from an economic and psychological point of view, the arrangements worked for her, in part because she did not have to report being divorced to her family, relatives, or friends. I have often thought about that interview, carried out with someone I did not know very well in the 1990s, because it was perhaps the first time I had heard someone speak so frankly and positively about in-house divorce.

2 Separate-surname activism

Oguni Kaori is the maiden name of a young woman who reluctantly changed her name when she got married in 2006. I had initially met Oguni-san after the hearing at the Tokyo District Court on October 10, 2012, some months into my ethnographic research on the *koseki*.[1] My friend Mikiko had emailed to alert me to the upcoming hearing and, although she could not attend with me, I was glad to see several people I knew in the courtroom that day. During the formal proceedings, Oguni-san gave measured but potent testimony on her experience changing her name to her husband's when they were married. The day I interviewed her in the summer of 2013, I wanted to go over some of the material she had presented in court back in October, but I also hoped to get her opinion about the future of the lawsuit and other *koseki* issues.

I met with Oguni-san on a warm July afternoon to discuss the lawsuit over the right of women to keep their own names in marriage. We chatted about the Tokyo humidity and wondered how much hotter the summer could get. I explained the ins and outs of the Institutional Review Board process at colleges and universities in the United States and we agreed that since she was a plaintiff in a visible lawsuit, which was in the courts, it would be best to use her real name in my work. When I pressed start on the recorder, Oguni-san grinned and said, "*jankenpon*" ("rock, paper, scissors"). She preceded to tell me about the night when she and her then fiancé began talking about the issue of their marital surname. "I brought up the subject and told him that one of us would have to change our name when we got married because of the Civil Code." "He said he knew." She smiled and straightened her skirt and took a sip of her tea before continuing. "I told him that I did not want to change my name." "When he said he didn't want to change his either, I said, how about *jankenpon*." Her admission that the couple used the traditional "rock, paper, scissors" game to decide a matter of such importance to her surprised me. Oguni-san made it clear during the interview that she had studied the matter of surnames carefully before she initiated the apparently light-hearted conversation with her husband. She knew she faced a dilemma, she said. Everyone expected her to take her husband's name and she really wanted to keep her own. Still, she went on, "I wanted to get him thinking about what it would mean for one of us to change our name and I was hoping to start a conversation. Instead, he sort of chortled and said, *jankenpon*?"

She turned away and then back to face me and in a serious and resigned voice said, "After that, we never discussed it again." In this part of the interview Oguni-san was explaining her attempt to engage a reluctant fiancé to at least consider her point of view. Instead, he laughed and dismissed the topic of conversation. Her description of the official name change was equally bleak, "On the day of the wedding we walked from the shrine in Asakusa to Taito District Office (Taito *kuyakusho*) to register our marriage. There I changed my name to my husband's name."

In her testimony in court Oguni-san talked about the discomfort she felt being called by a new name after she was married. She described a sense of disconnection when her "husband's name," as she called her new name, was called at the bank or the doctor's office. "Why are they addressing me by my husband's name?" she wondered. Her sense of disconnection was matched by a nostalgia for the person she had been and for the way people had addressed her for most of her life as Oguni-san or Ogu-chan (an endearment used by most of her close friends she said). Where was that person, she wondered? After a life of being called Oguni-san, or Ogu-chan, it was strange to hear someone else's name when she was addressed.

In the interview she repeated these things and said, "It feels like they are talking to someone else. Not me." "It isn't that I love my own name so much, but rather that my name feels natural, and I'm used to it." Oguni-san said:

> I'm sort of typically Japanese. Most people have addressed me by my last name my whole life. I know more people are using their first names recently, but I have never asked people to call me by my first name and most people do not.

When I asked Oguni-san why she changed her name if she was ultimately opposed to doing so, she said that she was worried about not being in a *koseki* with her husband. "What if something happens to one of us?" "There will be no proof of our relationship and we will have to use lawyers and 'prove' that we are a couple."

Of course, without a legal marriage a couple does not look like a couple at the level of the *koseki*. This points out one of the great benefits, and perhaps, the ongoing *raison d'être* of the *koseki* – namely, proof of kinship. When someone dies and a death notification is filed, everyone on the *koseki* of that individual is considered for inheritance and succession purposes. These days, families are increasingly small, but when older people die they tend to have a significant number of children, grandchildren, or other kin, legitimate or otherwise, who may be considered for inheritance because of the networks that can be gleaned from the main *koseki* or those of registered relatives.

On a related matter, most people do not write wills in Japan in part because the *koseki* can provide the evidence of legal heirs. More importantly, perhaps, is the tradition of single-child inheritance, so common until the 1940s, which made wills virtually unnecessary. Single-child inheritance, whether primogeniture or ultimogeniture allowed property and assets to stay together in the family name and at a particular location. The system allowed for a potentially smooth transition where in many cases the child to carry on the family name and property already lived on the premises and played the role of head of household whether a daughter or son of the deceased. When primogeniture or any single-child inheritance was replaced in the postwar period by a more "democratic" system that grants a portion of the property to the surviving spouse and treats all legitimate children equally, wills did not suddenly become popular because, again, there was a legal structure that could be used as a default in the absence of a will.[2] The relationship between legitimacy status, the *koseki*, and inheritance documents will be discussed more thoroughly in Chapter 3.

In Japan, the idea of "maiden name keeping," called *fūfubessei* (separate surnames for married couples) or *sentakuteki fūfubessei* (optional separate surnames for married couples) is a central feminist issue because of Article 750 of the 1948 Civil Code based on the 1898 Meiji Civil Code, which requires that upon marriage a couple must assume a single surname:

> 夫婦は、婚姻の際に定めるところに従い、夫又は妻の氏を称する。
>
> Upon marriage a couple shall assume either the husband's or the wife's surname.

The article means that either the husband or the wife must change their name. Separate surname activists have been trying to change Article 750 since the early 1990s so that it will be easier for those women who want to keep their maiden names in marriage to do so. Common practice and hegemonic norms lead to the expectation that a woman will willingly and graciously take her husband's name upon marriage unless there are unusual circumstances. Regarding this point, Oguni-san told me, "I'm Japanese, so I knew I would change my name." It is not impossible for a wife to keep her name, after all, in approximately 4 percent of all marriages, a husband assumes his wife's name upon marriage. However, this exception generally stems from strategic negotiations related to larger family needs or financial family dynamics rather than the insistence of a wife who wants to keep her maiden name.[3]

The surname is central to the patrilineal and patriarchal *koseki* structure, which in turn is central to the ideology or understandings of the family. The relationship between Civil Code 750 (the same-name rule), the *koseki* document, and ideas or ideology of family is made explicit in the marriage notification form required to become legally married. Below I discuss how this document constructs understandings of marriage and actual legal marriage. Most activists who critique the *koseki* critique the same-name rule of Civil Code 750 which is a central pillar of the *koseki* structure.

The Marriage Notification form (*kekkon todoke*) (Figure 2.1) specifically asks which surname will be taken by the marital couple. Of course, this is evidence that choosing the husband's surname is not a foregone conclusion. Information from the notification form is used to create a new or update an existing *koseki*. The sample Marriage Notification in Figure 2.1 is taken from the handbook, *Koseki jitsumuno tehiki* 2011, used in most government offices to guarantee compliance with the rules and laws surrounding the *koseki*.[4] It provides the legal codes backing each of 32 different status changes that may be made to a *koseki*. The book provides sample notification forms and sample altered *koseki* for each of these status changes, such as name change, marriage, and divorce. The Marriage Notification form must be completed and officially stamped using a *hanko* (personal stamp used instead of a signature for most official business in Japan)

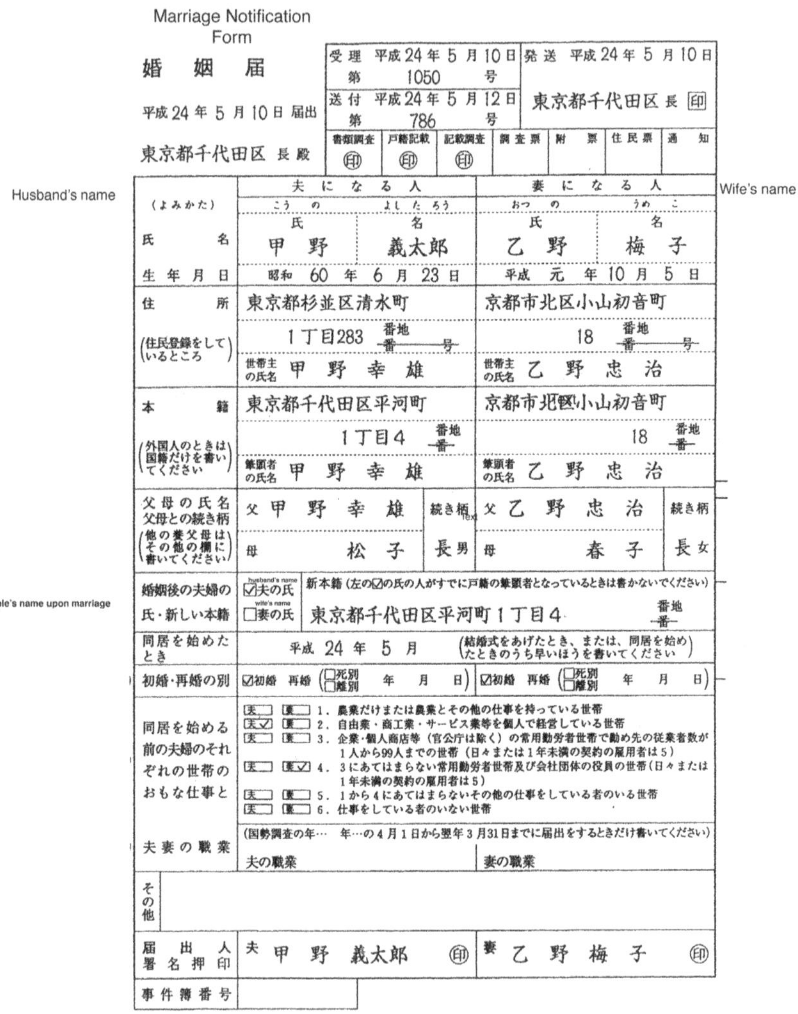

婚　姻　届

平成24年5月10日 届出

東京都千代田区 長 殿

受理 平成24年5月10日 第 1050 号	発送 平成24年5月10日
送付 平成24年5月12日 第 786 号	東京都千代田区 長 印

書類調査	戸籍記載	記載調査	調査票	附票	住民票	通知
印	印	印				

	夫になる人		妻になる人	
(よみかた)	こうの	よしたろう	おつの	うめこ
氏名	氏 甲野	名 義太郎	氏 乙野	名 梅子
生年月日	昭和60年6月23日		平成元年10月5日	
住所 (住民登録をしているところ)	東京都杉並区清水町 1丁目283 番地		京都市北区小山初音町 18 番地	
	世帯主の氏名 甲野幸雄		世帯主の氏名 乙野忠治	
本籍 (外国人のときは国籍だけを書いてください)	東京都千代田区平河町 1丁目4 番地		京都市北区小山初音町 18 番地	
	筆頭者の氏名 甲野幸雄		筆頭者の氏名 乙野忠治	
父母の氏名 父母との続き柄 (他の養父母はその他の欄に書いてください)	父 甲野幸雄	続き柄	父 乙野忠治	続き柄
	母 松子	長男	母 春子	長女
婚姻後の夫婦の氏・新しい本籍	☑夫の氏 □妻の氏	新本籍(左の☑の氏の人がすでに戸籍の筆頭者となっているときは書かないでください) 東京都千代田区平河町1丁目4 番地		
同居を始めたとき	平成24年5月 (結婚式をあげたとき、または、同居を始めたときのうち早いほうを書いてください)			
初婚・再婚の別	☑初婚 再婚 (□死別 □離別 年 月 日)		☑初婚 再婚 (□死別 □離別 年 月 日)	
同居を始める前の夫婦のそれぞれの世帯のおもな仕事と	夫□ 妻□ 1. 農業だけまたは農業とその他の仕事を持っている世帯 夫☑ 妻□ 2. 自由業・商工業・サービス業等を個人で経営している世帯 夫□ 妻□ 3. 企業・個人商店等(官公庁は除く)の常用勤労者世帯で勤め先の従業者数が1人から99人までの世帯(日々または1年未満の契約の雇用者は5) 夫□ 妻☑ 4. 3にあてはまらない常用勤労者世帯及び会社団体の役員の世帯(日々または1年未満の契約の雇用者は5) 夫□ 妻□ 5. 1から4にあてはまらないその他の仕事をしている者のいる世帯 夫□ 妻□ 6. 仕事をしている者のいない世帯			
夫妻の職業	(国勢調査の年… 年…の4月1日から翌年3月31日までに届出をするときだけ書いてください) 夫の職業		妻の職業	
その他				
届出人署名押印	夫 甲野義太郎 印		妻 乙野梅子 印	
事件簿番号				

Figure 2.1 Marriage Notification form.

in order to make a marriage legal. Each section on the form asks for particular information about the husband and the wife. In section (4) the marital name is required. In the left most box on that row is written, *Name of the couple once married*. In the next box to the right of that box there are two small check boxes and the words *Husband's Name* and below it *Wife's Name*. Only one name can be checked and as will be discussed later, one name must be checked. In the example, the check mark is in front of the character for husband, 夫. So the example shows that the husband's name will be used in marriage. Next to that box on the same row is a space for the new address for the couple, if there will be a new address.

The individual whose surname is chosen becomes the *hittōsha* (first registrant) on the *koseki* form. That person's name will be the surname for all members listed on the *koseki*. The physical *koseki* document, reflecting the Civil Code, allows only one surname. In the sample Parental *koseki* in Figure 2.2, read from right to left and top to bottom in traditional Japanese form, the domicile and surname are listed in the first column on the far right of the *koseki*. Each family member listed after (to the left) of the first registrant is named by given (first) name only because it is understood that all members listed share the same surname. It would be redundant to list the surname for each family member given the single surname principle. Here we can see that the structure of the *koseki* depends on a shared surname. In this sense, the *koseki* document both reflects and compels one surname in accordance with the legal constraints of Article 750 of the Japanese Civil Code.

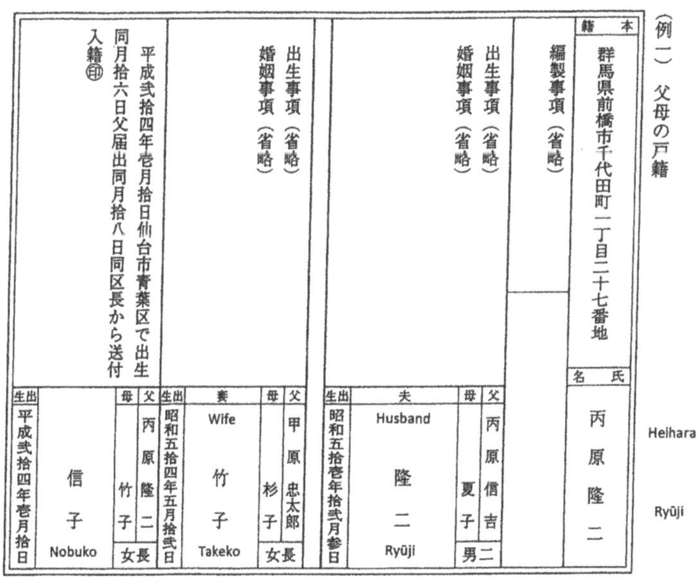
（例一）父母の戸籍

本籍 群馬県前橋市千代田町一丁目二十七番地

氏名 丙原隆二

Heihara

Ryūji

編製事項（省略）

出生事項（省略）

婚姻事項（省略）

父 丙原信吉 母 夏子 男二

夫 Husband 隆二 Ryūji

生出 昭和五拾壱年拾弐月参日

出生事項（省略）

婚姻事項（省略）

父 甲原忠太郎 母 杉子 女長

妻 Wife 竹子 Takeko

生出 昭和五拾四年五月拾弐日

平成弐拾四年壱月拾日仙台市青葉区で出生

同月拾六日父届出同月拾八日同区長から送付

入籍㊞

父 丙原隆二 母 竹子 女長

信子 Nobuko

生出 平成弐拾四年壱月拾日

Figure 2.2 Parental *koseki*.

In the approximately 3.6 percent of Japanese marriages in which the wife's name becomes the household name, tradition rather than feminist volition is generally at work. A practice that ensures continuity of a family line known as *mukoyōshi kekkon* 婿養子結婚 "marrying-in husband marriage," also understood as "son-in-law adoption," has existed for centuries in Japan when an eligible male heir is lacking. Son or son-in-law adoption to take over the head of household role through marriage to a daughter was chosen even in some cases when a son existed (Smith 1974). To improve the family fortune, augment the family lands, or for other generally strategically, economic reasons, a daughter might be kept and a husband brought in to take her name. In the early twenty-first century this practice still exists if great wealth or a prestigious family line/name are at stake, but another common reason to adopt for succession is so that there will be future descendants to care for the elderly and eventually the ancestors.[5]

Adoption for succession can be done in several different ways. A young man can be adopted as a son-in-law to marry the daughter of the house. In this rare case, his name would be changed to hers, but he would be the first registrant on the *koseki*. Another possibility is that a daughter would be the first registrant and a son-in-law would be married in as her spouse. In the past, some families lacking an eligible heir might even adopt an already married young couple into their family and into their *koseki*.[6] While son-in-law adoption seems to strengthen a woman's position since she would probably stay in her own natal home and inherit the resources of her parents, son-in-law adoption essentially maintains a patriline. The father's name continues through his daughter and adopted son and the blood/ethnicity/identity runs through the daughter for one generation. I will say more about this practice when we look at the changes in the *koseki* form from the Meiji period to the postwar period. In this chapter I examine the lawsuit challenging the same surname principle of Civil Code Article 750 and the related attempts to gain separate surnames by several generations of Japanese women.

Many Japanese lawmakers in the National Diet are opposed to changing the Civil Code, claiming that Article 750 is gender neutral because it states that either the husband's or wife's name can be chosen for the marital name and because of fear that women's independence regarding their names might destabilize the family structure. But *fūfubessei* activists argue that the well-established norm of choosing the husband's family name and the social expectations that coerce a wife to change her name lead to an inegalitarian system in which one's "choice" is highly curtailed by social practices and attitudes. In her study of American marital name-changing habits, the legal scholar Elizabeth Emens points out that even though American women have the legal right to keep their own names in marriage they tend not to do so in part due to social expectations and in part because their children are very likely to have their husband's name. Two dominant conditions in American naming practices, that men rarely change their name in marriage and that children are overwhelmingly given their father's names, means marrying women experience a more curtailed "choice" about their names than their husbands do. In the United States husbands almost never change their names in marriage, and as a result they experience continuity with their natal family, their wives (who change their names), and their children (who take their names). Emens emphasizes that women, on the other hand, very rarely have name continuity throughout their lives with children, husband, and their own parents. If they take the unconventional route of keeping their maiden name (still hovering around 10 percent of all US marriages) they are likely to have a different name than their own children (Emens 2007). In Japan, the framing of Article 750 as a matter of "choice" ignores the powerful social norms at work on women's naming options. Let me turn then to the separate surname lawsuit that was filed in Japanese district court by a group of lawyers on behalf of five plaintiffs of whom Oguni-san was one.

夫婦別姓 **(*fūfubessei*) separate surname lawsuit**

On February 14, 2011 a group of lawyers on behalf of five plaintiffs, four women and one man, filed a lawsuit against the Japanese government seeking damages for suffering caused by the loss of their surnames in marriage. The plaintiffs alleged that this loss was a direct result of a failure by Japanese legislators to create a law that would allow couples to maintain two separate surnames in a marriage. The lawsuit has been dubbed the Valentine's Day Case, because it was filed on February 14, 2011. It is also referred to as the *bessei* 別姓 (separate surnames) or 夫婦別姓 *fūfubessei* (married couples, separate surnames) case. This terminology reflects emic understandings of the surname issue in Japan. We might say in English a campaign to *keep a maiden name*, but Japanese does not reference the "maiden" name, a term rendered, but rarely used, as 旧姓 *kyūsei* (former surname). On the other hand, *bessei* literally means separate or different surnames and *fūfubessei* means couples with different surnames. The thing emphasized is keeping distinct or separate names within the normative *fūfu* couple marriage. Rather than focusing on the individual woman, or her name, the couple's relationship as a *fūfu* is highlighted.

According to several lawyers on the case, the 2011 lawsuit was filed as a last resort after repeated attempts to create either a change in the Civil Code through legislation in the National Diet or a concession within the Rules and Regulations that guide the specific content of the *koseki* document. Beginning in the early 1990s, several bills were introduced to permit separate surnames within legal marriage. Feminist activists had won numerous legal concessions after riding a wave of energy that began with the 1975–1985 UN Decade for Women. By the early 1990s, however, the Japanese Diet was facing a country and economy in turmoil, as the implications of the imploded real estate bubble sent shock waves through every part of Japan and stirred upset in party politics that left the conservative Liberal Democratic Party (LDP) out of power after a nearly 50-year term of control of the Diet. In the midst of this economic and political upheaval, women's issues and even the topic of women's equality in society became fodder for conservatives who were eager to find a scapegoat for their poor governance record.[7] After only three years of coalition governance by smaller and weaker political parties, the LDP was back in power with the election of Hashimoto Ryutaro in 1996. Instead of looking forward toward building a more transparent and democratic government and society, his government looked back nostalgically to "the good old days" and argued that changing attitudes toward women's empowerment in society had resulted in a weakened family structure, which could be blamed for economic and societal ills. Within several years the government turned back the gains feminists had made in the previous decade and by the late 1990s a backlash against feminism and women's equality in society was well established (Yamaguchi 2014).

The Liberal Democrats and their conservative supporters encouraged plans to increase the birth rate by paying women to produce more children. In many prefectures and cities policies were implemented to make cash payments to families upon the birth of each child with a graduated value placed on each additional child in order to encourage larger families at a time when support for working mothers was weak at best. Such government policy toward increasing the birth rate focused on financial compensation rather than much-needed daycare infrastructure and eradication of gender bias in the workplace which feminists argued would make it easier for women to both have children and stay in the workforce. By the time the newly formed Democratic Party of Japan upset the LDP in 2009 and took over the Diet, many women assumed the change to the single surname rule would finally come, nearly 15 years after the first legislation was introduced. However, political scandals and the devastating events of the Fukushima disaster on March 11, 2011 took precedence over everything else and women's issues, as they had been many times before, were tabled for what was considered more urgent matters of economic policy and recovery. Social and economic instability that resulted from the Fukushima disaster led to the re-election of the LDP by 2012 when the third Democratic Party prime minister in three years was ousted in a speedy election.[8]

Throughout the 1990s and early 2000s women's groups lobbied Diet members over the single-surname rule, and some members of the Diet had been keen to make the change (Tomizawa, 2013 interview). As the years went by and no solution was reached, however, lawyers and activists began talking about another tactic. According to Oguni-san, these conversations and the sense that the government lacked the will to change Civil Code 750 led to the Valentine's Day case filed in February 2011. The five plaintiffs were all in various study or activist groups trying to figure out how to keep their own names in marriage. Several lawyers who were working with one of them suggested that pooling several individuals into one case would help make a stronger lawsuit.

Sakakibara Fujiko, author of *Josei to Koseki: Fūfubessei jidai ni mukete*, (*Women and the Household Register: Toward an Era of Separate Surnames in Marriage*) is one of the lead lawyers on the case. I interviewed her in her office in an upscale part of Tokyo about the *fūfubessei* case in 2013. According to Sakakibara-san, the suit depends on Article 24 of the Japanese Constitution, which guarantees equality between husband and wife in marriage, and CEDAW, the UN Convention on the Elimination of all forms of Discrimination against Women, which Japan ratified in 1985.[9] These key documents support total equality for women, but as Sakakibara-san told me, they will not necessarily convince a Japanese court of law. The UN Committee on the Elimination of Discrimination against Women in its capacity to promote and ensure CEDAW compliance has urged the Japanese government to resolve the separate surname issue for married couples for many years.[10] Sakakibara-san told me that the current trend of over 96 percent of women taking their husband's name in marriage reflects traditional customs that are oppressive to women. Thus, in her view, it is not good enough that the Civil Code states that either the husband's or the wife's name can be used. Rather, to fully support Article 24 of the Constitution, the law must make it easier for women to keep their surnames in marriage, in other words, for both men and women to keep their own names.

When I interviewed Masami Kittaka, a junior lawyer on the case, she told me that the suit basically claims an infringement of the rights of the plaintiffs "for the government's failure to amend a law or make a law." The new law must provide a legal way for couples to marry and maintain separate surnames she said. Currently, the Civil Code makes this impossible. Either the Civil Code must be changed or a new law written. The plaintiffs' lawyers claim that the failure to pass a law providing this equal treatment to women infringes the rights of the plaintiffs.

In addition to Oguni-san, there are three other female plaintiffs and one male plaintiff in the separate surname lawsuit. Oguni-san told me that they were brought together by the lawyers, although some of them had met each other previously through the *bessei* activist network. Three of the plaintiffs are from the Tokyo area, one is from Osaka and one from Toyama, a western prefecture next to the Sea of Japan and a good four-hour train ride from Tokyo. Tsukamoto-san, a retired schoolteacher from Toyama, is the eldest plaintiff. She has three children and is currently legally married, in other words, she is registered in a *koseki* with her husband, under his surname. However, for much of their marriage, Tsukamoto-san was legally divorced in order to maintain her maiden name, Tsukamoto. Tsukamoto-san's testimony centered on her identity as a teacher and a mother and she described the trouble she has had trying to keep her own name through years of teaching and childrearing. Tsukamoto-san is not currently her legal name. However, since she is retired, and as she says, "no one can stop me from using it," she goes by that name.

Although many women who want to continue using their maiden names do so regardless of their legal marital names, not all employers will allow it. This was the problem for Tsukamoto-san who for many years had used her maiden name at work. To her students and colleagues she was known as Tsukamoto Sensei. (Sensei is a term of proper address and respect for teachers, doctors, and masters of traditional arts.) When policy at the school changed and everyone was required to use their legal surname for all school business, her principal told her that she would have to use the name on her *koseki* (her legal name) at work. If she had been registered under Tsukamoto, she was told, she could use that name.

In court she said, “I am Tsukamoto Sensei.” “I did not want to become ‘somebody else’ Sensei and to be called by a different name.”

But keeping her maiden name on paper (in the *koseki*) meant she would have to divorce her husband. Tsukamoto-san wanted both marriage and the use of her maiden name. She was willing to live in a common-law marriage and forgo the benefits of legal marriage in order to keep her name, but the situation was complicated because she had children with her husband. Common-law marriage, which for some activists results from paper divorce, is one way to keep a maiden name, but as these activists well know, common-law marriage can make a woman and her children vulnerable in Japanese society. Oguni-san told me about her sense that unmarried couples and their children are vulnerable if something should happen to one of them and they do not co-exist in one *koseki*. To a greater extent, single mothers and children born out of wedlock are in a vulnerable position in Japan because of centuries-old ideas that condone a man’s sexual liaisons with multiple mistresses and concubines, and condemn the mistress/concubine as whores and their offspring as bastards. The *koseki* has marked “illegitimacy” very explicitly until recently and children born out of wedlock have been considered the bastards of philandering men and irresponsible women. The term often used disparagingly for mothers of children born out of wedlock is “*kawaiisoo*” (pitiful). In addition to the condemnation of the women who have extramarital sexual relationships with men, a number of other things, including abortion and attitudes toward premarital sex, have helped to keep the rate of out-of-wedlock births at under 5 percent since the end of World War II.

In Japan, subtle, but persistent, discrimination against both children and their single mothers has been kept alive through the *koseki*. Unlike in the west, where in France, for example, approximately 50 percent of all children are born to unmarried couples or single women and in the United States, where the total rate of children born to single mothers is nearly 30 percent, Japanese women tend to avoid becoming single mothers if at all possible.[11] However, it is worth saying, that among feminist activists and other well-educated women, a small and smart population of women want children more than husbands. The Single Mothers’ Forum, introduced at the beginning of the book, has been supporting all stripes of single mothers for over three decades and long-time leader and activist, Chieko Akaishi, is confidently and happily a single mother by choice, refusing to resort to marrying the father of her child just to avoid discrimination or to make rearing her child go more smoothly.

Tsukamoto-san, however, from an earlier generation and a more rural and conservative part of Japan, had no intention of letting her children get registered on the *koseki* as born out of wedlock when she was essentially married (common-law) to her long-time partner. While still unmarried when her first child was born, she went to the local government office and submitted both a marriage notification and a birth notification simultaneously, rendering the child legitimate.[12] Soon after that, she divorced her husband on paper (*kami rikon*), by filing a divorce notification at the same government office, so that she could resume using her maiden name. The child stayed in the father’s *koseki*, noted as legitimate. When she gave birth to a second child, she did the same thing, re-registered the marriage to the same man (whom she considered her husband – she'd been living with him and raising a child with him) and registered the birth of her second child on the same day. Soon after that she submitted another divorce notification, a paper divorce, in order to legally maintain the name Tsukamoto. This complicated strategy for "keeping her maiden name" was worth it since she cared so deeply about the matter.

For Tsukamoto-san, Oguni-san, and Yoshi-san, all married plaintiffs in the Valentine's Day lawsuit, the legitimacy status of their children was an important factor in the choice of their marital and *koseki* statuses and, ultimately, their participation in the lawsuit. Tsukamoto-san's repeated paper divorces were a strategic way to guarantee that she was legally married when each child was born so that their birth notifications and *koseki* showed that they were legitimate. Only after the paperwork for the births had been completed (and the children were safely identified as legitimate) did she file for a divorce. For Tsukamoto-san this was a hassle that was only made tolerable because of the extent to which she wanted to maintain her maiden name. The paper divorce option made it possible for Tsukamoto-san to keep her maiden name, but she admitted in court that her husband did not like that they were technically divorced. Nonetheless he supported her decision and they stayed "happily married." It wasn't until after the birth of her third child that Tsukamoto-san was notified that the school did not approve of her unconventional marriage. She was told by the principal that the school would not recognize anything other than a legal marriage.[13] She was basically forced to register her marriage in the *koseki*, so that she was not an "unmarried mother." "Start using your husband's name beginning next semester," she was told. This development meant that in order to continue teaching (in the public school), she was being forced to take her husband's name and be registered as his wife in his *koseki*.

As a result of the change in school policy, she became increasingly frustrated and angry with the new policy, which was making it so difficult for her to do what seemed reasonable and natural, maintaining her professional name and identity while being married with children. For Tsukamoto-san the choice was now between her cherished job and her name. She had no intention of giving up her job and, thus, she ultimately stayed legally married and began to use her husband's name at work, unhappily. At the time, and still today, Tsukamoto-san said that she does not view herself as a troublemaker. She was willing to follow the rules even though she was terribly disappointed. It wasn't until after her retirement that lawyers and activists encouraged her to consider joining the lawsuit for the good of other women who are in a similar situation.

The third legally married plaintiff, Yoshii-san, like Oguni-san and Tsukamoto-san, is currently married and registered as a wife in a *koseki* headed by her husband. She is a home economics instructor at a Junior College in Osaka and the mother of two children. On October 10, during the hearing, her friendly husband sat holding their little baby girl who slept through much of the testimony. Yoshii-san said, "I had written my thesis and published several things under my name, Yoshii, so I did not want to change my name when we got married." "That is why we decided to have a common-law marriage." When one of the lawyers asked why she eventually did submit notification for a legal marriage, she said, "when it came to getting company housing through my husband's employer, they required us to bring our marriage notification, *kekkontodoke*."[14] Company housing is a common benefit with mid- to large-size companies and tends to be well under market prices, whether as a rental or through a purchasing program. The young couple did not feel they could forgo the housing benefit for the sake of separate surnames, so they filed a marriage notification. When asked how she felt after going through with this, Yoshii-san said "My husband is the eldest son in his family so we chose Tani as our name, but I was tormented by a very strong sense of loss." Despite her cheerful and charming presence, she said, "I became very sad."

The last two plaintiffs in the *fūfubessei* case are a couple in a common-law marriage who were formerly legally married in one *koseki*. Mr. Watanabe holds an engineering job and Ms. Kayama works as a freelance writer. Both testified in court about the inconvenience and emotional difficulty of name changing. Ms. Kayama told the court that she was happy when she married Mr. Watanabe and understood that her legal name would be Watanabe. She assumed it would be easy to continue to use her maiden name like many writers do. However, something started to occur not long after she got married. Although she had been known by the name Kayama, she was increasingly questioned about the use of her name, Kayama, when all of her legal documents said her name was Watanabe. In court, she said, "in numerous situations I could not use 'Kayama' even though it is my name. This started to feel strange," she went on, "like a fictional name, a borrowed thing. And eventually I felt that my actual existence/*sonzai* had become like a borrowed thing *karimono*, in other words, not my own."

In his testimony, Mr. Watanabe said he understood his wife's distress and frustration. However, he said, "I didn't want to change my name either." "I had become used to my own name, and I have a special license for my work registered under my name." "When we went to register our marriage at the Arakawa District office in early 2011, we tried to register our marriage with both names." Figure 2.1 shows a Marriage Notification form and, as discussed earlier, only one surname can be chosen. Ms. Kayama and Mr. Watanabe chose both names by putting a check mark in front of his name and her name, and submitted the form at the desk. However, their notification was rejected and their marriage was not accepted. After that they filed a lawsuit in Tokyo District Court against the Arakawa ku government for damages for not being able to marry and maintain separate surnames. However, within weeks the case was dismissed and they were told that the complaint should be handled at the Family Civil Court *katei-saiban*.[15] The dismissal of Kayama and Watanabe's first suit fueled the decision to sue the national government for failing to change the law in 2011. Mr. Watanabe explained during his testimony in court that in addition to the many inconveniences and the sense of insecurity he and Ms. Kayama feel by not being "legally married," there is also an economic disadvantage to a common-law marriage. When the couple applied for a mortgage recently, it was turned down because they were unmarried, even though they had ample income for the loan.

After giving their statements about the inconveniences of having to use the same surname in marriage at the October 2012 hearing, the plaintiffs were called back to court on May 15, 2013 for a long-awaited decision. I attended this court session and felt palpable excitement in the room that day. After so many years of waiting for some public authority to state officially that women deserve both a *de jure* and de facto right to keep their name in marriage, there was a sense of hope. If the court did not have something positive to offer, at least my informants assumed it would point the way forward to the next step toward realizing dual-surname marriages. Instead, in a weak voice, Judge Masako Ishiguri gave the court's decision, "the Constitution does not guarantee a couple's right to use separate surnames" she said. Furthermore, she stated, it was not in the power of the court to change a law. "We cannot do anything," she said. After the one-minute decision was read, the judge solemnly said "the case is dismissed" and everyone filed out of the courtroom. Out in the hallway after the decision was announced, several lawyers made brief comments on the decision and invited everyone to a meeting room downstairs where the legal team reported on the decision and tried to make sense of the court's action. Many of the people in the courtroom who had been involved with the case volunteered to lobby legislators, write op-ed pieces for newspapers, and give other support to couples seeking the right to dual surnames. The mood in the room, though crestfallen, was optimistic and energetic. All the lawyers seemed convinced that their work on separate surnames must continue.

It was almost two years later, in March 2015, that the Supreme Court of Japan announced that it would consider the case of separate surnames, along with a case claiming discrimination against women because of the rule of re-marriage after divorce (the Civil Code required that a woman wait a minimum of six months after divorce to remarry; there is no such time requirement for men). In December 2015 the decision on the two interrelated cases was handed down. The court decided against awarding damages on the *fūfubessei* case, they could not make a statement on the right of couples to maintain separate surnames and said this is work that must be done in the National Diet. The related case on the waiting period to remarry after divorce was won by the plaintiffs and the change was made effective immediately. The defeat of the *fūfubessei* case is just one more set-back in a nearly three-decade battle for separate surnames in marriage in Japan.

During this long period in Japan some women determined to keep their maiden names have never legally married their partners. These individuals, in long-term, monogamous, marriage-like relationships, decided that living as married without legal marriage gave them the right to keep their names and afforded other kinds of independence from traditional marriage. In the next chapter we will look at long-term common-law marriage as a strategy to circumvent the same-surname rule and the constraints of the *koseki*. Tanaka Sumiko has been an outspoken advocate of non-marriage and has published extensively on her own experience. Along with her long-time partner, Fukukita, she has buoyed a movement of marriage resisters who challenge all aspects of the male-headed family represented by the *koseki*. Marriage resisters have been able to keep their own names, but their non-marital status has forced them to face other entanglements with the *koseki*, the most pressing of which is the fact that their children are considered "illegitimate" by law. These issues and the *koseki* implications of non-conformity with the patrilineal surname rule will be examined through the experiences of Tanaka and several other *koseki* activists.

Notes

1 In the interview with Oguni-san, I referred to her as Oguni-san, her maiden name, which she prefers to use even though it is not her legal name. I refer to each person in the book differently reflecting the nature of my relationship with them and often their preferred term of address. Context, nature of relationship, and age of interviewee, among other variables, determine what naming practice we will use. Closer friends generally call me Linda and I call them by their first names, including Mikiko, Saeko, Yumiko, and others. I introduce myself as Linda and make it easy for people to address me by my first name, and people I refer to with last name and san tend to refer to me as Linda-san, in part because they are well aware of US conventions where addressing people on a first name basis is widespread. I believe many of us negotiate terms of address in the field depending on a wide range of factors. In Japan, this is complicated by subtle norms that do not necessarily hold with non-Japanese. Another important factor in my relationships relates to feminism and the extent to which people I work with and interview know European and North American practices. In feminist and activist circles I am associated with in Japan, informality and a resistance to hierarchy and prestige status informs choices about naming. But regardless of my location in Japan, being addressed by my first name helps me develop the rapport I try to create in my fieldwork relationships. Ms. White or Professor White would not be conducive to the relationships I'm trying to build with feminists and other interlocutors. As an American, I'm very comfortable being addressed by my first name in Japan and I try to sensitively read interviewees for clues as to what is appropriate. There is no need to ask. If someone doesn't suggest I call them by a first name, I follow the dominant cultural pattern and use the last name as I did with Oguni-san and many other people who helped with my research.

2 Beginning in 1947 surviving spouses were entitled to one-third of the property and assets of their deceased spouse and the rest was divided among surviving legitimate children, with children born out of wedlock entitled to one-half of the amount due to legitimate children. However, single-child inheritance by consensus seems to have continued due to the practice of one child caring for and living near elderly parents. In the 1980s the legal division was changed to grant one-half of all property and assets to the surviving spouse and the rest for the children, continuing the differential between legitimate and "illegitimate" children which continued to be a source of activism until 2013. In a landmark Supreme Court decision in 2013, the differential apportionment of inheritance to children born out of wedlock was determined to be unconstitutional, so now one-half of property and assets is assigned to the surviving spouse and the rest is divided *equally* among all children (legitimate and illegitimate) of the deceased. Of course, it is not always simple to figure out who the children of the deceased are, because children born out of wedlock are not on the *koseki* of their fathers. If a child does not have a relationship with his or her father, he or she is unlikely to make the claim for inheritance. This will be further discussed in Chapter 3.

3 Records of various types of husband adoption or son-in-law adoption being used historically in Japan show that a male marries into the wife's household/name/*koseki*.

Often referred to as *muko yōshii* (婿養子), where *muko* means husband or groom and *yōshi* means adoption. In such cases the husband is adopted into the wife's family and generally becomes head of the household. Nonetheless, in such cases a wife would be able to keep her surname and usually live in her natal home in marriage.

4 I am indebted to Tanaka Sumiko who gave me her personal copy of the handbook, *Koseki jitsumuno tehiki* 2011, that she had acquired after years of employment at the Musashino City Hall. I have since discovered that the book is available on Amazon Japan.

5 One of the central tenets of filial piety in the Japanese family is care of the elderly and the deceased through ritual observances such as memorializations discussed in Chapter 1. The *koseki* is one site of locating filial relationships and obligations.

6 In his fine study of ancestor worship, Robert J. Smith situates the *mukoyōshi* pattern of succession with a discussion of five different ways in which succession has been guaranteed in Japan historically.

7 See Yamaguchi (2014) and Wakakuwa and Fujimura-Fanselow (2011) for a recap and analysis of some of the most devastating aspects of this attack on gender equality and social justice campaigns.

8 This is not the place for a lengthy discussion of postwar political structure in Japan, but it should be noted that conservative control of the government for most of the postwar period has negatively impacted not only the separate surname campaign, but women's equality in all realms of society, despite Prime Minister Abe Shinzo's claims to care and promote women, known as "womenomics." Since the current Imperial Reign began in 1989, Japan has had 16 different prime ministers, several of whom, like Prime Minister Abe, have been at the helm of the party and government more than once. The short run of the more progressive Democratic Party of Japan included three prime ministers between 2009 and 2012.

9 Article 24, which guarantees equality between husband and wife in marriage, along with Article 14, which states that there shall be no discrimination based on race, creed, sex, social status, or family origin, are the two cornerstones of women's legal rights in Japan today and are attributed to the active role played by the multi-lingual, 23-year-old Beate Sirota Gordon who worked closely with the Allied Occupation in 1946 to write Japan's postwar Constitution and pushed for these basic human rights for women (see Gordon 1997).

10 The draft report for the October 2015 CEDAW session provides a list of 17 issues for the Japanese government to continue addressing, including five related to violence against women, three related to women's health, and one focused on "marriage and family relations." The issue raised here encourages Japan to "introduce a system that allows a husband and wife to adopt separate surnames" (CEDAW Draft, Sixty-third Session Report, Advance Unedited Version, July 30, 2015).

11 See Hertog (2009) for a thorough analysis of this issue.

12 Legitimacy status does not seem to correlate to date of conception. Registration of both marriage and birth on the same day made the child legitimate and this practice of registering both on the same day was not uncommon among my informants who were actively maintaining their maiden names.

13 A full discussion of this decision was not had in court. What right did the school have to tell Tsukamoto-san that she could not get a paper divorce in order to use her maiden name? Under paper divorce, she was technically in a common-law marriage and perhaps this was perceived as unseemly for a schoolteacher. It is not clear what standing the school had to require her to use her married name. Given the circumstances, I think we can assume that common-law marriage, even if motivated by a desire to maintain a maiden name, is perceived as unacceptable for someone in the position of school teacher. Social pressure rather than legal right may have been at work here.

14 Yoshii-san said "marriage notification" in court, though it is unlikely that an employer would ask for the notification (which is the paperwork completed by the individual and submitted to the *koseki* desk at the local government office in order for a *koseki* to

be created or altered). Generally, notifications are not requested by employers, banks, private detectives, and so forth. Rather, what is required is a copy of the *koseki* or a copy of a section of the *koseki*. A *jūminhyō* (resident register) a parallel registration, which pertains to a residence rather than kin relations captured on the *koseki*, contains much of the same information, since it is all generated from the same set of notification forms it is also frequently used in situations like housing, loans, and employment. The *jūminhyō* (resident register) figures centrally in the lawsuits against the government discussed in the next chapter.

15 This information comes from the website of the *Fūfubessei* Lawsuit Support Group. www.asahi-net.or.jp/~dv3m-ymsk/tusin.html. The group formed in 2011.

3 Common-law marriage as a form of *koseki* resistance

For women, all children are legitimate (*Onna ni totte, Kodomo wa subete, "chakushutsushi"*).

Tomizawa Yoshiko

Tanaka Sumiko is a dedicated and determined *koseki* activist, well known among lawyers and activists working to change the Civil Code and Koseki Law for her quiet, persistent intensity, and for her good humor and wit. She has made the issue of discrimination against children born out of wedlock her life's work, despite having a full-time job as a civil servant, since she bore her daughter out of wedlock in 1985. Tanaka was determined to keep her name and stay single when she became pregnant, leading to her confrontation with established registration rules and laws. I first met Tanaka on a sunny fall Sunday in 1994 at a rally for the rights of children born out of wedlock. Tanaka and her partner, Fukukita Noboru, were gathered with a group of other activists in a park near Shibuya station in western Tokyo, where they were making signs, talking, laughing, and enjoying the day. Shibuya, famously crowded with shoppers, families, and young people on Sundays, was an ideal spot to be seen and heard by people out for a good time and not in a particular hurry. Unlike rallies I had attended in the United States against nuclear power, war, or sexual assault, this rally was called a *pareido*, パレード parade, and had the feel of celebration and carnival rather than resistance and social critique. Fittingly, people were dressed in bright colors, music was playing, and the scene was joyful.[1] Some of the activists carried banners that said *Kodomo o mamorō*! (Let's protect children!) and *Kongaishi mo byōdō desu*! (Children born out of wedlock are equal too!), and *Kodomo wa minna byōdō*! (All children are equal!). My friend, Barbara Summerhawk, herself a friend of Tanaka Sumiko, had invited me to join the event and introduced me to Tanaka and Fukukita that day. As we strolled through the crowded neighborhood with the approximately 50 others who had gathered, I saw the surprised looks on shoppers' faces. Yes, they seemed to say, yes, children should be treated equally, but why should children born out of wedlock be treated the same as legitimate children? After all, the distinction provides the basis for protection of the institution of marriage. It is widely believed in Japan that children born out of wedlock are inferior to legitimate children (Hertog 2009).

Discrimination against children born out of wedlock has been naturalized in modern Japan. The ubiquity of heterosexual marriage and the prohibition of out-of-wedlock birth (not sex) is in line with a generous attitude toward abortion and men's extramarital liaisons. As men experience great latitude in understandings of their sexual practices, women as mothers or as non-marital lovers shoulder the burden of guaranteeing that children are born within wedlock. Historically, women have used many strategies to avoid giving birth to a child out of wedlock. In the twentieth century abortion was used to great effect to keep the rate of illegitimacy extremely low. In recent decades, shotgun marriage as well as abortion have helped keep the out-of-wedlock birth rate below 3 percent. (Hertog 2009). Although Tanaka Sumiko did not consider herself a single mother nor view her child as one born out of wedlock, the Koseki Law and Civil Code specify that a child born to an unmarried woman could not be registered as legitimate.

In the introduction to an anthology on the rights of children born out of wedlock that Tanaka published in 2005, she writes:

> even before I had considered the possibility of having children, people would chastise me as a "bad mother" and lament my children's situation, "*hidoi hahaoya da*" (what a terrible mother) and "*kodomo ga kawaisō*" (the pitiful child) for not registering my marriage.

She goes on, "it made no sense given that I had not even considered having children yet."

Article 49 of the Koseki Law states that notification of the birth of a child must be made within 14 days of birth.[2] The ritual of registering a child's birth is significant and is mentioned often in the stories my informants tell. Tanaka and Fukukita, soon after their daughter was born, went to register the baby's birth at the local municipal office in Musashino City. There they faced the first set of challenges to the registration of their child, the Birth Notification form required for the legal registration of a child upon birth. Figure 3.1 is a sample of a Birth Notification form required by the Koseki Law and Civil Code.[3] Near the top of the form I have highlighted the section for legitimacy status of the child. There are two small boxes in front of the terms for "legitimate" *chakushutsushi* 嫡出子 and "illegitimate" *chakushutsu de naiko* 嫡出でない子. In the sample, the top box for "legitimate" is checked. When they submitted their Birth Notification form for their daughter, Tanaka and Fukukita simply left the boxes blank. Tanaka told me during one of our early interviews in 1995 that from their point of view, their child was not born out of wedlock (she used the term *kongaishi* 婚外子 as she always does, because it is less pejorative than the official term, *chakushutsu de naiko* or *hichakushutsu*) since they had been living together for ten years when she was born. However, because they had no marital *koseki* on file at the city office to prove their relationship they were not allowed to check "legitimate" on the form.[4]

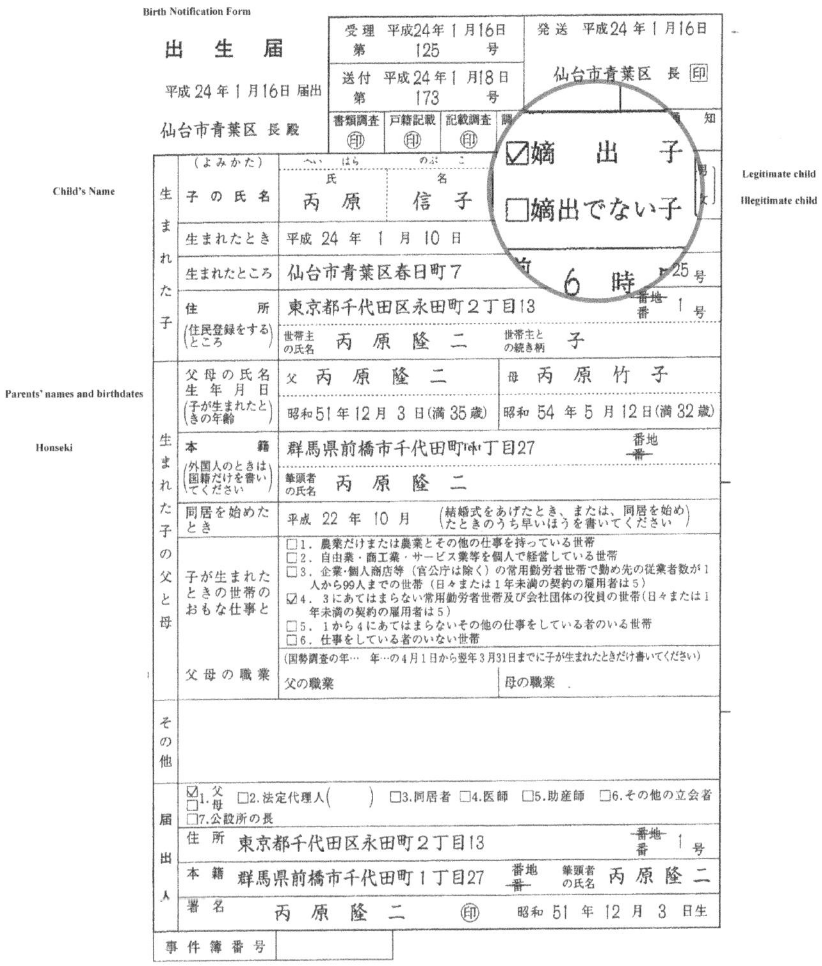

Figure 3.1 Birth Notification form.

When the couple tried to submit the Birth Notification form without checking either the box for legitimate or the box for illegitimate, the official at the *koseki* registration desk asked them to wait while the Legal Affairs office, part of the national government under the domain of the Justice Ministry, was contacted. Tanaka-san told me that they waited for over two hours that first day, trying to get permission to submit the Birth Notification without identifying the child as either legitimate or illegitimate. Ultimately, after their long wait, the form was accepted without either box checked. Nonetheless, when the *koseki* was issued some days later, it labeled their daughter as born out of wedlock. Eager to have the child registered as legitimate they began consulting with people at the city government, Musashino City, where they both happened to be employed as civil servants.

Tanaka and Fukukita consulted with many people before they filed their lawsuit against the government. They requested an official negotiation with the Mayor of Musashino City and other city officials to have their case given special consideration because they had been living together as a couple for a very long time. They hoped they would be able to convince people who knew them and their relationship that an exception could be made. Tanaka told me on more than one occasion that a lawsuit was, for them, a last resort, not a first option. In fact, they did not file suit until their daughter was already three years old.

Since the early postwar period citizens have been registered in not only a *koseki* by family relationship, but also in a register called *jūminhyō*, which tracks households through residence. The problems of rural–urban migration that have plagued *koseki* documentation since the Meiji period are somewhat alleviated by the parallel *jūminhyō* register that reflects where people live (not their permanent domicile which is a central organizing principle of the *koseki*). Records of *jūminhyō* are maintained at the city, village, or county office where people actually live. Addresses on *jūminhyō* thus must be within those districts. *Koseki*, on the other hand, are maintained in the district where a permanent address is registered. For many urban Japanese, the *koseki* has been maintained in a rural area where grandparents or great-grandparents originally lived. Until recently, urban Japanese returned to their hometowns to carry out *koseki* notifications and to obtain copies of their *koseki* for things like school applications, marriage negotiations, or proof of identity. In the postwar period the *jūminhyō* has been substituted in many cases when proof of citizenship, marital, or other status is required.

For Tanaka and Fukukita, the discriminatory designation of their daughter was, of course, on both registers and they ultimately decided to begin their lawsuits upon the advice of lawyers with the *jūminhyō*. Thus the first lawsuit against the Mayor of Musashino City was for discrimination in the residence register *jūminhyō*. Tanaka-san confirmed with me during a conversation in 2013 that they had been advised by several lawyers to initially address discrimination in the *jūminhyō* because it was perceived as somewhat less symbolic of family and tradition in Japan and there might be a higher likelihood of winning a case of discrimination in this less culturally potent registration system.

The representation of illegitimacy on the Birth Notification form is straightforward with one of two boxes checked. That information is then used to specify a child's status on the *koseki* and *jūminhyō*. On every entry for every person listed on a *koseki* there is a section where the gender and birth order is listed, known as *tsuzukigara* 続柄. The *tsuzukigara* 続柄 is essentially a marker of ranking for succession within the family. For children born out of wedlock, a different way of designating them on the *tsuzukigara* immediately reveals that they were born out of wedlock. Thus, the first lawsuit for discrimination against their daughter in the *jūminhyō* pinpointed the *tsuzukigara* as the main site of discrimination. Both the *jūminhyō* and the *koseki* clearly identified their daughter as born out of wedlock. In Figure 3.2, I have labeled the *tsuzukigara* column on the sample *koseki* from the Koseki Manual (2011: 43). For legitimate children the characters for gender and child are written as they are in this sample.

On the *koseki*, children's location in the family is identified using special terms that rank children by gender and birth order, eldest daughter, *chōjo*, eldest son, *chōnan*, second daughter, *nijo*, second son, *jinan*, and so forth, in the *tsuzukigara*. These terms are also used in informal conversations when referring to a person's family location and birth order, referencing the *koseki* where they are officially documented, reflecting the continuing significance of birth order and the *koseki* itself as meaningful signifiers of identity and status within the family in Japan today.[5] But as Tanaka-san and Fukukita-san pointed out, these special birth-order terms were used strictly for the children of legally married parents. For children whose parents were not registered together in a marital *koseki* they could not be used. Instead, for such children of unmarried parents, just gender, using the characters for female (女 *onna*) or male (男 *otoko*) was used on the *koseki*. On the *jūminhyō* children born to legally married parents were registered with a combination of gender and birth order as on the koseki. Children born out of wedlock, however, were noted with gender and the term child, showing immediately that they were not ranked by birth order as legitimate children. Tanaka-san told me that anyone could quickly ascertain that a child was born out of wedlock by the absence of the specific terms for legitimate child that are widespread and normative in the culture.

Claims for damages due to discrimination against their child in a lawsuit first filed against Musashino City in 1988 did not lead to a victory in the courts

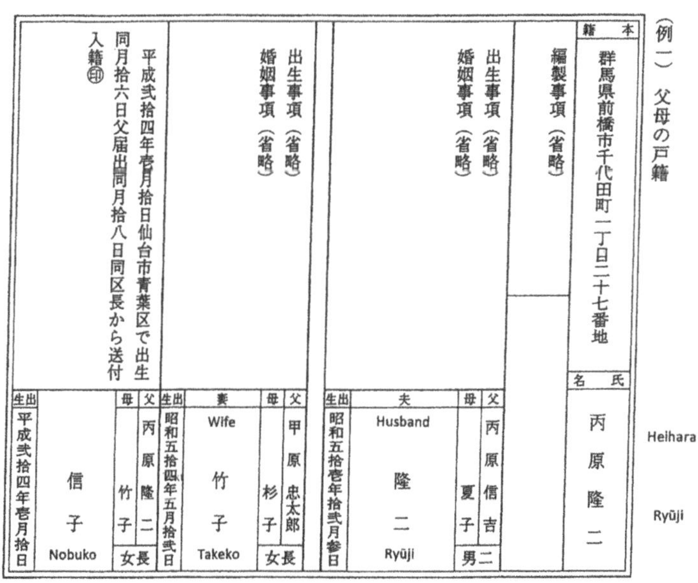

Figure 3.2 Parental *koseki*.

(Tanaka-san and Fukukita-san lost their case and several subsequent appeals). After these setbacks Tanaka-san and her team of lawyers and supporters, then called the Association in Support of the Lawsuit Against Discriminatory Expressions in the Resident Register, filed a petition addressed to the Minister of Justice, the Chairman of the House of Representatives and the Chairman of the House of Councillors requesting changes to the Civil Code and Family Registration law to eradicate discriminatory references to children born out of wedlock (See Appendix I for a brief list of key activities of the Association [*kōryūkai*].) In 1992, feeling frustrated by the insular nature of the Japanese court system, the Association in Support of the Lawsuit Against Discriminatory Expressions in the Resident Register decided to send a message to the United Nations Committee on Human Rights, submitting a Counter Report Against the Report of the Japanese Government concerning the International Covenant on Civil and Political Rights to the Human Rights Committee of the United Nations, dated July 20, 1992. In the report the group explains the discriminatory notations on both the *jūminhyō* and *koseki* and asks for the committee's consideration. Later that fall, Tanaka-san and Fukukita-san went to Geneva to personally present their Counter Report and to observe the discussion of their claims. Tanaka wrote about that experience in her 2004 anthology, describing the sense of hope she and Fukukita gained from presenting the material to the UN committee in an international forum (Tanaka 2004b).

During the next few years the support group organized lectures, lobbying trips to Diet members' offices, and negotiations with lawyers. In 1993 they held a large gathering with members of five other groups working to address the treatment of children born out of wedlock and discrimination in the *koseki* and *jūminhyō*. Then in an unexpected shift in policy, in December 1994, the Ministry of Home Affairs (*Sōmusho*), the bureau of the government responsible for management of the *jūminhyō* system, announced that henceforth notations for all births on the *jūminhyō* would be the same, listing the gender, but not the birth order, of each child, stating that the distinction between illegitimate and legitimate children would no longer be made on the *jūminhyō* document. Tanaka and Fukukita, although they had lost their appeals in the Tokyo courts, were vindicated when the change was put through from the government bureaucracy. The Ministry of Home Affairs' administrative change, conceding the claims of the couple and their supporters, made any further appeals on the discrimination case moot. The court convened again in March of 1995 to inform Tanaka and Fukukita that their appeals on the case for discrimination against their daughter were no longer "applicable." At the same time, the administrative move abolished a stubborn relic of the old regime and ultimately the discriminatory distinction on the *jūminhyō* was eradicated.

A similar resolution came after the couple sued the government for discrimination against their daughter in the *koseki* in a subsequent case that lasted from 1999 to 2005. After appealing an adverse lower court decision to the Tokyo High Court, and losing at that level in 2004, the Legal Affairs Bureau of the Ministry of Justice announced in 2005 that the notation for all children, regardless of the marital status of their parents, would be consistent, the gender/birth order terms on the *tsuzukigara* of eldest daughter (長女 *chōjo*), eldest son (長男 *chōnan*), and so forth, which had been used only for legitimate children up until that time, would now be used for all children. Many years of hard work and political organizing had finally paid off. Beginning with their early negotiations with Musashino City in 1985 and their ongoing lawsuits and appeals, the couple had amassed an incredible support system that included *pro bono* legal aid from the likes of Fukushima Mizuho and Sakakibara Fujiko, among others. And they had gathered an incredible amount of information about the *koseki* and *jūminhyō* system. I think it is fair to say that only lawyers and academics who have studied the two systems for years know as much as Tanaka Sumiko. In addition to her own self-styled education about the Civil Code, Koseki Law and the minutiae of *koseki* protocol, she has advocated for other single mothers and their children, educating many activists who subscribed to her *mini-komi* journal, *VOICE*, which she continues to put out monthly or bi-monthly with a small group of supporters. Collating and labeling an issue of *VOICE* one afternoon near Kichijoji in western Tokyo with her, Tanaka-san told me that the group of supporters who helped her lobby, produce *VOICE*, and ultimately continue the lawsuits had made it possible for her to keep going all these years. It wasn't easy, she said with a full-time job and a daughter to raise, but friends and supporters made it worthwhile. "We just couldn't give up."[6]

The celebrations of the changes to the *tsuzukigara* had barely started when it became clear to Tanaka-san and Fukukita-san that the registers would continue to be a site of discrimination for children born out of wedlock. It was true that children born after the 2005 decision would not have the blatant blemish of illegitimacy on the *tsuzukigara* section of their *koseki*, but for children like their daughter, the *koseki* would need to be changed to eradicate the discriminatory notation. Tanaka-san changed their daughter's *koseki* and *jūminhyō* as soon as it was possible to do so, but as with all changes to these documents, evidence of the change is manifest in an explanation in the margin next to that individual's entry. In other words, changing the child's designation would just move the information (about out-of-wedlock birth) to another location on the *koseki* and *jūminhyō*. From the time of their first lawsuit until today Tanaka and Fukukita have pursued a political agenda to eradicate discrimination in the *koseki* system. They worked tirelessly from 1988 to 2005 on their own lawsuits. Their activism has brought a hegemonic social structure that perpetuates privilege for men and those who are legally married into the light of day, helping to change people's consciousness about the social norms that shape their lives.

Essentially, Tanaka and Fukukita brought charges of discrimination against the state, spent years advocating their position that children should not suffer for the decisions of their parents, lost their cases, but ended up providing the impetus for the system to be changed by fiat from above. Most people give credit to Tanaka Sumiko and Fukukita Noboru, and their lawyers and supporters, for the changes in how out-of-wedlock children are registered on the *koseki* and *jūminhyō* today. But while the most egregious forms of discriminatory notation against children born out of wedlock on the *koseki* and *jūminhyō* were abolished, there are still a number of ways in which illegitimacy can be identified in both documents. Importantly, the Birth Notification form, a notification form which signals a change to an existing *koseki*, still specifically requires the filer of the notification to identify a child as legitimate or illegitimate. Thus, discrimination against children born out of wedlock on the Birth Notification form is a related and ongoing cause for Tanaka and Fukukita as well as other activists trying to eradicate the last traces of discriminatory practices related to the *koseki*, as I will discuss below.

Although the explicit demarcation of illegitimacy which appeared in the *tsuzukigara* column on the *koseki* and *jūminhyō* has been removed for all newly made *koseki* as a result of the changes made by the government after the years of legal pressure from Tanaka and Fukukita and their supporters, many older *koseki* that have not been updated show the status of children as illegitimate, because there is no automatic updating of *koseki*. Until a family member comes to make a change by filing a notification form, family documents will not be altered. As mentioned earlier, when Tanaka Sumiko submitted her Birth Notification for her daughter in 1985 she did not want to check either box regarding legitimacy because she resisted the distinction. At the time the city allowed her to submit the form without checking either box after consulting with higher authorities. Activists challenging discrimination in the Birth Notification form today are no longer able to submit a form without checking one or the other box. They have tried to submit the form without checking the box for legitimate or illegitimate but their notifications have been rejected as a result.

The Birth Notification and the strategy of non-registration of one's child

One such activist, a father committed to equality between men and women and children born out of wedlock, Sugawara Kazuhiko, has been particularly critical of the *koseki* system and the unfair treatment of women and children in the *koseki*. He and his partner tried to register the birth of their daughter without checking the box for *chakushutsu de nai shi* (illegitimate). However, the Birth Notification form was not accepted and they left the city hall much disheartened and unsure of what to do next. After consulting with some of the same activists involved in Tanaka-san's case, he and his partner, who is not named in any lawsuit in order to protect her privacy, decided to not register the birth of their daughter at all in order to avoid complying with the options that would have forced them to name her as *chakushutsu de nai shi* (illegitimate) on the notification form. As a result, the couple's daughter was not registered on either parent's *koseki* and technically lacked citizenship and official proof of identity.

Sugawara filed suit against his local government for discrimination against his daughter on the Birth Notification form on the grounds that although the *koseki* and *jūminhyō* documents no longer distinguish between legitimacy and illegitimacy, the Birth Notification form, required for any *koseki* alteration, cannot be filed without specifying the legitimacy status of the child. The entanglement of *koseki* and notification forms means that it is not enough to say that one's child is legitimate, although one must in order to register the child that way, but the parents who complete the notification must be legally married, share a common surname, and have a *koseki* as a couple. This is why the problems of the legitimacy/illegitimacy distinction and separate-surname issue are inextricably linked.

Similarly, although the *tsuzukigara* column of the *koseki* where birth order and gender are listed is no longer the site that reveals illegitimacy there are other ways to ascertain that a child has been born to unmarried parents. (The primary way to ascertain out-of-wedlock birth is the registration of the child on a *koseki* headed by its mother, a single mother, with no spouse.) The deep social opprobrium for children born out of wedlock will be considered further in the following chapter. Underpinning the activist work to change the notation of such children is the desire to expose and exorcise this remnant of Japan's sexual double standard, which sanctions children born within marriage, and excludes those born to concubines, mistresses, or other sexual partners. The legal experts Sakakibara and Fukushima as well as plaintiffs in cases over the treatment of children born out of wedlock show how the *koseki* system privileges and protects men and their legal wives while discriminating against the children and non-marital mothers who have them (Sakakibara 1992; Fukushima 1992; Tanaka 1995).[7] Below I will introduce Tomizawa Yoshiko, a key activist who resisted legal marriage in order to keep her surname and kept her only child off of her *koseki* in order to avoid the discriminatory notation of illegitimacy.

> To women, all children are legitimate (*Onna ni totte, Kodomo wa subete, "chakushutsushi"*).
>
> (Tomizawa Yoshiko)

Soon after I arrived in Tokyo in 2012 to study *koseki* activism, Yumiko Jansson-Yanagisawa, an old and beloved friend whom I first met in the 1980s and a long-time feminist activist, called and invited me to come to her place in central Tokyo to meet Tomizawa Yoshiko, a friend and *koseki* activist, whom Yumiko said could help me with my research. Yumiko and Tomizawa-san went back a long way to at least the 1980s when they were both active in that period's heady feminist movement, and although Yumiko and I knew many of the same people by now, I had never met Tomizawa-san. Yumiko generously invited the two of us for a Sunday afternoon meal and discussion in mid-October, where we discussed Tomizawa-san's rather unique approach to *koseki* registration that had started over 30 years earlier, before any changes had been made to the registration system brought about by the lawsuits and international lobbying of Tanaka-san and Fukukita-san and other *koseki* activists.

Tomizawa-san, like Yumiko, is well educated, politically engaged, and energetic. Yumiko wrote a column that often took up trending gender issues for *MORE* magazine in the 1980s and 1990s and she is also an accomplished translator of English and Swedish literature, having translated Annie Dillard, Alice Walker, and Nadine Gordimer among other important novelists who work in English. Tomizawa-san has worked as a community activist who often takes up the cause of gender inequality in Japan and has contributed to many *mini-komi* journals, the backbone of Japanese feminism, over the past 30 years. From the 1980s well into the 1990s she served on the Suginami Ku local government board, re-elected numerous times. Tomizawa-san became an expert on the *koseki* through her own long-term struggle to protect her son from discrimination and to keep her maiden name.

When I formally interviewed Tomizawa-san not far from her home in Suginami-ku several months after I met her, she told me that she wanted to keep her name and her sense of identity when the decision about her child and marriage arose in 1983. When Tomizawa-san became pregnant while she was living with her boyfriend, a still fairly unusual arrangement in those days, she made a conscious decision that greatly shaped her life and ultimately her son's as well. Tomizawa-san decided to stay unmarried even once her son was born. Registering the marriage would require her to take her partner's name and conform to an old-fashioned notion of family that she resisted.[8] She said she was happy with her relationship with her partner and saw no need to legally sanctify it through *koseki* registration.

Per law and custom, Tomizawa-san and her partner had gone to register the birth of her son within 14 days of his birth. As she described what happened that day, it seemed starkly vivid in her mind as she recounted the event of nearly 30 years ago:

> I submitted the Birth Notification form that day and it was accepted at the counter. I did not check either box related to legitimacy. I did not consider my son illegitimate and I refused to put a check in front of that box. Neither was the child legitimate, so I left that box blank also. Everything seemed fine until I was contacted by the city office several days later and told that it would be impossible to process (*uketorenai*) the Birth Notification if I did not specify the legitimacy status of our son.

Her eyes narrowed with indignation as she continued:

> I was told to return to the office to complete the registration by completing the form; in other words, they would not accept the registration if I did not label my son born out of wedlock. But, I did not consider my child "born-out-of-wedlock" since I was in a relationship, although not a legal marriage, very much like a legal marriage. I refused to check the box that said "*hichakushutsu*" (illegitimate), so, in the end, our son was unregistered until he was 11 years old.[9]
>
> (Interview, October 28, 2012)

During the decade when her child was unregistered and lacking citizenship and legal identity in Japan, Tomizawa-san and her partner had to make special requests to gain access to daycare and primary public education and to obtain mandatory childhood vaccinations and other benefits of national medical care, which they ultimately did. All these privileges and services become available to citizens and in most cases, even non-citizen residents, through their *koseki* or *jūminhyō* registration, she told me.

Tomizawa-san and other informants told me that in the 1980s a copy of the child's section of the *koseki* was generally required with an application to enter school.[10] Without the *koseki*, the child might be assumed to be non-Japanese (since all Japanese are normally registered in their (Japanese) parents' *koseki* and *jūminhyō*). Without this proof of birth and family affiliation individuals need other proof of identity. For foreigners the Alien Registration card was required from 1952 through 2012 when a new system requiring long-term foreigners (over six months) to register in the *jūminhyō* system was introduced.[11] It is worth noting that until 2012 resident foreigners were excluded from both the *koseki* and *jūminhyō*. The change in 2012 has important implications particularly for long-term Zainichi residents of Japan, most of whom were born in Japan but do not have citizenship.[12] Given the number of foreigners living, working, and supporting families in Japan it is high time that they have some way to engage with the institutions that support and permit health insurance, daycare, and public education.

It should be clear that Tomizawa-san and Fujita-san have a very unique situation. They left their child unregistered for many years to resist and critique the system that would call their son "illegitimate." Tomizawa told me that she and her partner were focused on two specific things when they initially went to make the birth notification. They wanted to avoid the negative categorization of "illegitimacy" so they did not complete that part of the Birth Notification form. They also wanted their child to be registered in his father's *koseki* and thus have his father's surname. Article 790, 2 of the Civil Code, states, "An illegitimate child assumes the surname of its mother," and Article 52 of the Koseki Law states, "A notification of birth of an illegitimate child shall be given by its mother." These two articles determine that a man cannot register his illegitimate child in his own register, nor can he bequeath his name to that child. Furthermore, only a head of household or a head of household in cooperation with a legal spouse can make entries into a *koseki*. This is why children born out of wedlock remain in the *koseki* of their mothers. This is one of the fundamental ways in which out-of-wedlock births are recognized as such in Japan today. By definition, a *koseki* where a child born out of wedlock is registered will be headed by the child's mother, not its father. Thus, although a mother may try to protect her out-of-wedlock child from discrimination, it is not very difficult for the child's status to be discovered. Tomizawa-san's request to have her child registered in the father's *koseki* subverted the very logic of the *koseki*.

The question of the child's surname also goes to the heart of *koseki* logic. Although *koseki* activists have for the most part been concerned about their own maiden names, the issue of children's names comes up in cases of divorce, and for Tomizawa and Fujita it was an issue because they particularly wanted their child to have his father's name. Because the Civil Code specifies that children born out of wedlock must be registered in their mother's *koseki*, it forecloses the possibility of registration in the father's *koseki* under the father's surname. Permission to register a child born out of wedlock in the father's *koseki* appeared absurd within a system defined by exclusive marital patrilineality.

Throughout our conversation that first fall afternoon at Yumiko's house, Tomizawa and I talked about children, motherhood, feminism, and the frustratingly slow pace of change in gender attitudes in Japan and the United States. When we circled back to the *koseki* issues, Tomizawa-san told me that when her son was about to enter Junior High School, she and Fujita-san went again to the Legal Affairs division of their local ward government in Suginami-ku where they lived to request permission to register the child on Fujita-san's *koseki*. This time, the person in charge said that she would take the request to the Justice Ministry (*homusho*), the bureau responsible for such *koseki* and Civil Code related issues. But again, after some time, the couple were told that it would be impossible to register the child on the father's *koseki*. As a result of that refusal and because the couple felt increased pressure to register a now older child who increasingly needed proof of identification, including date of birth, citizenship, and other vital information, later that same year they took their request to register the child to Family Court (*katei saiban*) in Suginami ku, this time pleading for the need to have the child registered. Tomizawa-san told me that "We told the local court that we had lived as a family for these eleven years and that our son had been unregistered because we did not want him to be registered as born out of wedlock." In a surprising turn, Suginami ku family court decided, despite the law and the precedent of previous decisions, to grant the request. Tomizawa told me that the judge said that because they had lived as a family together and because Fujita did not have any other children from another marriage for whom there might be an inheritance-related conflict, they would allow him to register his son, as eldest son with the name Fujita. "Since Fujita is not married to someone else, since he doesn't have other children and there won't be competition among children, we will allow this registration." The outcome was celebrated, and for Tomizawa-san it was a significant victory.

When I asked Tomizawa-san about the decision, she said, in retrospect, she was not sure why the Family Court finally decided in their favor, but added that perhaps they just wanted to get rid of a troublemaker. This interpretation is hard to believe given the long-term stubborn response of the state and its subsidiary local branches in town and village government to any changes in the *koseki* system. Calling herself a troublemaker was also in line with her self-deprecating tone which downplayed her significant role in getting the results that she wanted through patience and strategy. In fact, she might have said that perseverance, strategic timing, and stubborn demands for justice on her part had brought about the long-awaited outcome. Tomizawa's activism, her use of *mini-komi* newsletters and her support of other women in common-law relationships whose children were born out of wedlock made her a leader and spokeswoman for the campaign for equality for women in marriage and society. Her requests and discussion of her situation in the media brought bad press to the government, the *koseki*, and the male-dominated status quo in Japan. But despite her small victory, most couples who hope to avoid the discriminatory distinction made on the Birth Notification form opt to legally marry rather than leave their child off of the *koseki*, because of the many inconveniences and fears that their child's lack of proof of identity will have dire consequences at some future moment, as Oguni-san, plaintiff in the separate-surname case (Chapter 2) told me when I interviewed her.

The only other unmarried couple in my research group who left their child off of the *koseki* altogether in resistance to the term illegitimacy on the Birth Notification form are Sugawara Kazuhiko and his wife, discussed earlier in the chapter who also reside in Suginami-ku in western Tokyo. One of their children was unregistered for many years because they refused to select either of the options regarding legitimacy on the Birth Notification form. They also worked through Suginami-ku officials and the family court. Finally, in 2013, the recently elected mayor of Suginami-ku, Mr. Kosaka, gave permission to the local city *koseki* division to create a *koseki* and *jūminhyō* for their daughter, even though they did not ultimately check either of the boxes for legitimacy status on the Birth Notification form, because he was eager to see their case resolved. Sugawara and his partner were able to have the daughter's name entered on the mother's *koseki* without naming her as born out of wedlock on the Birth Notification, a small victory, but nonetheless, a victory.[13]

During 2012–2013 Sugawara-san gave several public lectures about his experience fighting the registration system. He is a soft-spoken, quiet man in his early forties. At his presentation at Nerima-ku City Hall he talked about the many conversations he had had with city officials regarding the discriminatory language on the Birth Notification. It was clear that like Tanaka Sumiko he saw his personal challenge to the *koseki* system as an opportunity to teach others about how the registration system works. He said that he tried to explain to local officials why he and his partner refused to register the child as "illegitimate." To the couple, their daughter was never an out-of-wedlock baby. But as Sugawara told me when I interviewed him, the cost of not registering a child can be high. If something should happen to the parents or the child her/himself the child is extremely vulnerable without proof of identity. Activists perceive their choice to leave their children unregistered as a great risk, which might have negative social and financial implications. Thus, non-registration of a child as a rejection of *koseki* discrimination is a highly unusual strategy. So too is suing the government over discrimination against children born out of wedlock. The parents who have taken these extreme steps to expose the contradictions of the *koseki* system challenge patrilineal *koseki* logic and heteronormative societal norms, risking their own reputations and peace of mind, not to mention the security of their children. Both Sugawara-san and Tomizawa-san have devoted a great deal of time to the *koseki*. The fact that they went so far as to not register their own children, whose well-being would clearly be served through registration, is significant. As has been stated previously, proof of citizenship, family relation, age, and so forth comes through *koseki* or *jūminhyō* registration. Without these documents one truly does not have a legal existence in Japan today.[14]

Tomizawa-san has been an active critic of the *koseki* system since her son's registration. Through several decades she has championed and supported other activists who have challenged the *koseki* and either sued the government or critiqued the system in other ways. She continues to fight to end discrimination against children born out of wedlock and to support groups working to improve the conditions of single mothers, both true single mothers and women in common-law marriages who appear single under the law. But ultimately, she does not identify as a single mother. She has never actually lived as a single mother and she has known financial and social security with her partner Fujita in their common-law marriage, *jijitsukon*. Unfortunately, her activism on behalf of her son and others like him did not change any laws or *koseki* rules. Instead, the Legal Affairs Bureau of the Justice Ministry granted an exception for her son that ultimately left the system and its biases against single mothers and their children in place.

After I met Tomizawa in the fall of 2012 her name came up frequently when I talked with other activists and insiders to the *koseki* lawsuits. She was known by many of my other informants as a strong, intelligent activist, dedicated to changing the system of discrimination against children born out of wedlock. Indeed, I went to several hearings with her in 2012 and 2013, and witnessed her distinction in the activist community by the respect with which others treated her.

Dual registries mean dual sites of discrimination

The dual system of *koseki* and *jūminhyō* is worth further discussion here. To some extent there is a redundancy of information in the two separate registration systems, but their purposes in modern society are quite different. The *koseki*, always translated as either household register or family register, has the longer history in Japanese society, evolving as it has from the early record-keeping system brought from China in the seventh century. Those early records, written on wooden tablets and indexed by specified pieces of land, or what we now understand as an address, recorded family members in order of entrance into the family. A variation of those early tablets shows up in the record-keeping system ordered by the Tokugawa shogun in Buddhist temples in the seventeenth century. At that time the documents provided information for local Daimyo (feudal lords) as well as the Shogun and provided a record of births, deaths, marriages, and divorces by household. According to Hayami and Cornell, through shogunal decree each family group was required to register at the closest Buddhist temple, informing authorities of births, deaths, marriages, and divorces. Each Daimyo and ultimately the Shogun had access to the information and used it to assess population, land use and yield, and a variety of other things. A great weakness of that Tokugawa-era registry, however, was that when people moved from one territory to another there was no consistent way to keep track of them. Thus, registries included names of people who had long ago left a particular area. This weakness in the registration system would become of critical importance by the modern era when vast numbers of people left the small villages and rural towns of Japan to head for education and jobs in larger towns and cities. The Meiji government's effort to count the population and monitor changes across the land was unruly. Indeed, the modern *koseki* system did not have a method for handling movement of family members from the original registered domicile.

The twentieth-century *jūminhyō* was designed in the early postwar period to be used by officials of local district governments, just as the earlier *koseki* had been used, to ascertain aspects of the population, such as number of school-age children, age and name of head of household, and so forth. Because the *koseki* was increasingly located somewhere other than where family members resided (due to ongoing urban migration) and because family members were dispersed across vast areas, the *jūminhyō* became the way to track current residence for various municipal services and records.

Related to the different roles they play, the *jūminhyō* and *koseki* function under different governmental authorities. According to Article 1 of the Koseki Law, "The affairs concerning family register shall be managed by the mayor of a city, town or village." But ultimately, the affairs of *koseki* are under the auspices of the Ministry of Justice while the *jūminhyō* is under the control of the Ministry of Home Affairs. Under a revision to government policy made in 2009 and in effect from 2012 all foreign residents of Japan, including all permanent, non-citizen residents of Japan and short-term residents and researchers, must now register in a *jūminhyō*. Including non-citizens in the *jūminhyō* marks another distinction between the *jūminhyō* and the *koseki*.

From an activist's point of view, the *jūminhyō*, like the *koseki*, provides a site of discrimination and differentiation. For many couples in common-law relationships the discrimination against their children is felt first at the level of the *jūminhyō* because their children's eligibility for education, healthcare, and other social benefits stems from their address and resources allocated from their local city hall. This is a further reason that Tanaka Sumiko and Fukukita-san first took up discrimination against their daughter in the *jūminhyō* before the *koseki*.

Tanaka-san and Fukukita-san found in their first lawsuit against Musashino City over discrimination against their daughter in the *jūminhyō* that the two systems of registration are distinct. Although differential treatment in the *jūminhyō* for children born out of wedlock was abolished in the 1994 decision, it took another ten years using the court system, the media, and the international sphere of the United Nations, to successfully convince the Japanese government to make the parallel change to end overt discrimination against children born out of wedlock on the *tsuzukigara* column in the *koseki*.[15]

Conclusion

Although Tanaka Sumiko is in so many ways a unique activist, she has many of the characteristics of other *koseki* activists. She, like them, is a low-key, relatively soft-spoken person. She talks about her own travails with the *koseki* quite calmly and methodically. As she recounts the events that shaped her activism, she tells of the relatively simple things she asked for, the ability to complete a notification form without misrepresenting the status of her child, the ability to consider her child as normal and legitimate. She was surprised by the defensive and rigid response she received from her employer and local government and assumed that an agreement could be made. But when her request was not granted she did not do what was expected, namely register her marriage so that her child would be registered as legitimate. Instead, in a very deliberate and determined manner, she did what she felt was right. She went to the supervisor of the person in charge of the *koseki* division who had declined her request and asked for a reconsideration of the policy. When that didn't work, she wrote letters to several representatives in local government – all subtle and straightforward processes to get what she believed was fair. But when permission to designate her child neutrally without the discriminatory designation of illegitimate child was denied, she and Fukukita-san decided to sue the government for discrimination.

In this chapter I have identified several different strategies for challenging the registration systems. Tomizawa-san and Sugawara-san and their partners ended up leaving their children off of the *koseki* for many years to avoid discrimination against their children. Tanaka-san and Fukukita-san, whose child was registered as illegitimate against their will, tried everything including attending a UN session of the Committee on Human Rights in Geneva to make their case for the dignity of children, regardless of their parents' marital status. Ultimately, Tanaka and Fukukita sued the government until the changes they sought were made. Still the most common response to the problem of illegitimacy of children born to couples in common-law marriage is a quick marriage followed by a paper divorce, a strategy used by many of the people I have interviewed.

Some of the activists discussed here reject everything about the marriage system and even find the idea of a common-law marriage objectionable. Others are concerned primarily with the right to separate surnames and for them non-registration of marriage in order to keep their name is enough. *In toto*, these activists have challenged the status quo through myriad unconventional practices to avoid changing their names, and they have forced the authorities in local government and at the national level to hear the complaints of women who find the *koseki* antiquarian and discriminatory toward women.

When children are present, the stakes of non-conformity to the registration rules are much higher, and the tabooed, and in their cases, inaccurate designation of "single motherhood" has fueled much of the activism against the *koseki* rules. It is worth noting again the incredibly hard work carried out by those who do not conform to the registration process like Tomizawa-san and Sugawara-san. They see their own children treated as social inferiors because their parents are not legally married; they see the bias toward single mothers (not truly single, but perceived as single) and they have had to advocate at many levels of bureaucracy during the early years of their children's lives. For Tanaka-san and Fukukita-san, who ultimately registered their daughter as born out of wedlock, the lawsuits and ongoing participation in demos/parades, lobbying, and writing about the *koseki* and the critical legal cases became a huge undertaking that has dominated over 30 years of their lives. In the next chapter, the underlying logic of the patrilineal *koseki* system will be examined in an effort to illuminate the slow pace of change to this deeply embedded social structure.

Notes

1 Clearly, protest varies around the world and since this early observation of a protest that struck me as whimsical and light for an issue that was clearly taken very seriously by participants, I've noted a major shift in the use of humor, color, and playfulness in protests in the United States. Perhaps some events like Gay Pride marches, now Pride, have always tended toward lighter and upbeat celebrations, but at the time I found the positive energy a welcome surprise.

2 More time is allowed if the child is born out of the country, but this three-month extension is notable for its relative brevity.

3 The Birth Notification form is printed in a manual called *Koseki Jitsumu*, created for city workers to explain the proper completion of such forms for the *koseki*. I gratefully accepted receipt of my copy of the 2011 edition from Tanaka Sumiko.

4 Emens and others discuss the problem of proof of relationship for long-term common-law marriages in the United States.

5 See Smith (1974) and Ueno (1987) for further discussion of birth order and gender hierarchy in the Japanese family. In fact, reference to sons or daughters could easily be through their family position as eldest daughter, *chōjo*, second daughter, *nijo*, or third daughter, *sanjo*, taking the place of personal names.

6 See Steinhoff (2014) for a discussion of the important role of support groups in long-term legal battles.

7 I am particularly grateful to Yumiko Jansson-Yanagisawa and Tomizawa Yoshiko for a long and fruitful conversation about the inherently conservative uses of the *koseki* to protect married women while condemning and excluding unmarried sexual partners of married men.

8 I learned a good deal about Tomizawa's situation that first day at Yumiko's house, but much of this story comes from a formal interview I did with her some months later in Ogikubo.

9 Tomizawa-san chose the term "*hichakushutsu*," the one used at the time on the Birth Notification form. Today the term used to designate out-of-wedlock birth is *chakushutsu de nai shi*. There is a distinction worthy of consideration. *chakushutsu de nai shi* can be translated as "a child born out of wedlock," while *hichakushutsu* does not have the word child in it, so has a gruffer tone that corresponds perhaps to "out of wedlock" or "illegitimate." Neither of these bureaucratic and technical sounding terms has the neutral and politically sensitive tenor of *kongaishi*, a term adopted in the movement against discrimination in the *koseki* that reflects a nuanced non-discriminatory tone.

10 There are several different ways to get a copy of a *koseki* from the city office. A family member has a right to request his or her own *koseki*. Often a school or employer will request a section of the *koseki* known as the *koseki shōhon* that deals specifically with the individual involved. Sometimes the entire *koseki*, the *koseki tohon*, is requested.

11 The inclusion of non-Japanese citizens in the *jūminhyō* system is worthy of further attention for several reasons including the fact that foreigners are so specifically marginalized in the *koseki*.

12 See Chapman (2011) for an excellent discussion of the inclusion of long-term, non-Japanese residents in the Resident Registration system. I was excited to have my

first encounter as a resident who needed a Resident Registration under the new system in 2012 when I arrived for my research on the *koseki* system after hearing about it for so many years from Tanaka-san and others. Unlike the koseki, the *jūminhyō* may have more than one surname, so it was possible for my daughter and me to be listed together.

13 Unlike Tomizawa-san's case they were not trying to get the child registered on his father's *koseki*, they just did not want to be forced to label the child *chakushutsu de nai shi* ("illegitimate") on the Birth Notification. The *koseki* would ultimately reveal that the child was born out of wedlock because the child would be registered in the mother's *koseki*, but they would not be coerced to classify her as "illegitimate" on the form.

14 While this seems entirely true, there is change afoot. During the past ten years a system called Jukinet and another called *mai nambā* have been established. In order to be in Jukinet one must be registered within *koseki* and *jūminhyō*, but *mai nambā* may circumvent the *koseki* system and make it possible to be registered there without a *koseki*. A recent Mainichi newspaper article discusses the recent proposal to link the *koseki* system with the *mai nambā* system, which would make what is considered private *koseki* information much more accessible to a range of state and local government parties.

15 See White (2014) for a lengthier discussion of Tanaka Sumiko's use of the international organizations and media to bring about an end to discrimination of children born out of wedlock on the *jūminhyō*.

4 Illegitimacy and male privilege

The underlying logic of the *koseki*

I met with Sumie Oshino, a charming and outgoing woman in her early fifties, a few weeks after my research assistant and friend, Saeko, introduced us through cellphone texts and then email.[1] Saeko had told me that Sumie's *koseki* changes were quite complicated and suggested I interview her. She mentioned that in recent years Sumie had started teaching women's self-defense classes after training in a number of different arts for many years. Sumie's particular experiences and her interpretation of them, that reveal some of the gender constraints of the *koseki* system, unfolded in our discussion over lunch a few weeks later. We met at the cafeteria on the ground floor of Crayonhouse, the famed and beloved women's and children's bookstore and meeting place founded by feminist activist Ochiai Keiko in the 1980s, which has served as a hub for the women's community in Tokyo for over 30 years.[2] As we tucked into organic and home-made traditional dishes, such as *nimono* (savory simmered vegetables like daikon and carrots) and barley tea, Sumie started by talking about her childhood. "We lived with my grandmother when I was young," she started. "Both of my parents had jobs so my grandma, my mother's mom, really raised me."

"My father was a member of Soka Gakkai" she said. "He was one of those serious ones, unlike my mother who did not care for the new religion." Soka Gakkai, the largest of the new religions, developed from Nichiren Buddhist roots in the 1930s and now has millions of adherents around the world. She said that he rose daily at 4:00 a.m. to pray and wanted Sumie to attend church with him. Her mother on the other hand was not enamored of the new religion and Sumie's maternal grandmother was downright opposed to it. Sumie talked about her grandmother with affection, saying she had tried to make life at home good for her beloved granddaughter. The main tension in the house, according to Sumie, was related to Soka Gakkai. Sumie said that when she was in elementary school her father became romantically involved with a woman at the church. Around the same time her mother started seeing another man also. Sumie went on outings with her mother and the male friend while her parents were still married and then when she was 11 years old her parents divorced. She told me that there was some discussion at the time about whether Sumie would live with her mother or father, but with the grandmother's pull, it was decided that she would stay with her mother and grandmother.

Soon after the divorce was finalized, Sumie's mother married her suitor, Mr. Nomura. For Sumie this event was tainted with trauma for no sooner had her mother begun living with her stepfather (they moved into his house) than he began to seek out Sumie in the evenings and molest her. She said that he would come into the room where she was watching television while her mother was making dinner and "put his hands all over me." Sumie described a grim situation where her stepfather had forbidden her to tell anyone. She assumed her mother and grandmother did not know what was going on. Now, many years later, she believes that both her mother and grandmother knew what was happening, but failed to intervene. Explaining this, she told me that her grandmother even put a small lock on her bedroom door, telling Sumie, "use it if your stepfather bothers you."[3] It was also her grandmother who encouraged her to marry at a young age so that she could get away from home. Indeed, Sumie did marry soon after graduating from high school. She married an eldest son of the Kimura household rather far from her natal home. The Kimuras had adopted her husband in the *mukoyōshi* fashion, to head the household, a practice common in the absence of a male heir, discussed in Chapter 2. The family expected the adopted son to bring in a daughter-in-law so that the couple's children could carry on the family line. Sumie did indeed have a son to carry on that family line, but she was unhappy in her marriage and ended up divorcing her husband. Admitting that she had married in part to escape her mother and stepfather, she said, there were problems in the marriage as soon as her son was born. When I tried to get details about her marital problems, she said that her husband treated her terribly, insulting her body and her behavior in front of his family members. She told me that she could not handle the way he treated her and spoke to her; "I could not take his disrespect for me. I divorced my husband and took my son with me," stymying the hopes of succession for the Kimura line. Perhaps it was because she had moved households and *koseki* multiple times, perhaps the abuse from a mother's second husband or perhaps the sense that her mother who should have protected her had betrayed her and ignored her daughter's silent pleas, but Sumie was highly critical of the social expectations she had felt as a daughter and wife. Perhaps all these things and more made Sumie critical of the family system and the *koseki* itself.

Upon divorce, Sumie, as a single mother and head of household, went to live in Tokyo. As a result of filing a Divorce Notification she started a new *koseki* of which she was *hittōsha* for her son and herself. She was now on her fourth *koseki* and reverted to her maiden name. As her family status changed through the years, the changes were reflected on the *koseki*. First registered as daughter to her biological father and mother, then moved to the *koseki* of her stepfather after he married her mother, she moved again upon marriage to a new marital *koseki* with her husband Kimura and then when she divorced she had to leave the Kimura *koseki* and create a new *koseki* which she headed. Sumie's identity changed as she moved through these different *koseki*. In her biological parents' *koseki* she was identified as eldest daughter, *chōjo* in the *tsuzukigara* column. When her mother was divorced, she and her mother were crossed off of her father's *koseki* and moved to her stepfather Mr. Nomura's *koseki*. On her *tsuzukigara* on the Nomura-headed *koseki* she was no longer eldest daughter, but rather adopted child because she was not technically Mr. Nomura's daughter.[4] When Sumie told me about her adopted status, she said, "Isn't it strange that I became an adopted child? This shows the impact of the male perspective (*otoko no shiten kara*) in the *koseki*."

Sumie's third *koseki* was headed by her husband Mr. Kimura upon their marriage. In the Kimura *koseki*, her husband was an adopted son and Sumie would have been the second registrant on that *koseki* as his wife. When she divorced Kimura she started a new *koseki* with her son. Her only alternative at that point would have been to return to either her mother's *koseki* with the abusing stepfather or to return to her own father's *koseki*. As a grown woman with a child she did not want to do either of those things. By the time I met Sumie she had married again and had a new *koseki*, number five, with Mr. Oshino, her second husband with whom she had another child. During the interview Sumie returned to a discussion of her changed designation on her stepfather's *koseki* as adopted. "I am not an adopted child to my mother," she said, "but because the *koseki* is oriented from the perspective of the male head of household *koshu no shiten kara da kara*, it makes it appear as if I were adopted with no blood relationship to my own mother." Here again she uses the term *koshu*, an older term, but one still relevant for the old-fashioned ideas she is referencing. Her stepfather was like a Meiji-era head of household, a patriarch, a *koshu*.

Her comment about her stepfather resonated with others I had heard about the structuring role of the head of household on the *koseki*. Tanaka Sumiko, among others, had emphasized that while the *tsuzukigara* column may appear to designate a child's relationship to his or her parents, in fact, it truly represents only the relationship between the head of the household, the *koshu* or *hittōsha*, male or female, but usually male, and each person listed on the *koseki*. By virtue of naming the *hittōsha*, first registrant, formerly the *koshu*, ahead of other family members, all other members are listed in relation to the first individual listed, as spouse of the *hittōsha*, eldest son of the *hittōsha*, and so forth. In the terms used and the tangible representations of these relationships on the material form of the *koseki* we can see that although the head of household is now understood as only the first registrant on a document, the true meaning of the first registrant is that that individual is the one that matters, her/usually his name is the official name of the family and represents the family unit by heading the *koseki* document. This is amplified by men's resistance to giving up their role as head of household and the government's resistance to considering alternatives, such as households headed by two people rather than one, or two names rather than one, or individual registries that would allow separate surnames. The distinction between eldest daughter and adopted daughter may seem minor, but for Sumie it shed light on a complex power structure within which she was a minor player, a pawn, first as one man's child and then another's, then one man's wife, and then another's, all the while her mother also changing her name and her *koseki* to match that of her husband.

A key objective of the postwar Occupation of Japan was to change the family culture which was blamed for strict hierarchy and requisite obedience to authority, key elements of Japan's long war in Asia. Eliminating the all-powerful role of the head of household was part of the Occupation's agenda to create equality in the family, as well as in society at large, and this was orchestrated by the new articles to the 1947 Constitution guaranteeing equal rights to all and equal rights to husbands and wives in marriage (articles 14 and 24 respectively). But despite these ambitious and admirable changes, the *koseki* and the household relations it represented continued in the postwar period, undermining the democratizing agenda of the drafters of the postwar Constitution. Equality between husband and wife as an ideal to strive for was touted with excitement by progressives in Japan in the postwar period, but the *koseki* document and Koseki Law and Civil Code were not altered, making the hierarchical rendering of the family on paper and in law fundamentally unchanged (Satō 1996).[5] The significance of the head of household, a critical component of the Meiji-era *ie seido* that was purportedly eradicated after World War II when Occupation lawyers worked to democratize the Japanese family and bring equality to men and women in society, continued under a new name.

Sumie critiques the male-household-head centrism of the *koseki* system that positioned her identity as permeable and deeply linked to a male family head, even when her biological mother was the link to the family. This male-household-head perspective is also at the heart of the illegitimacy issue discussed in Chapter 3. It was Tomizawa Yoshiko who said, "for women, all children are legitimate," pointing to a fundamental problem with the concept of illegitimacy. Tomizawa-san's *koan*-like comment asks us to ponder what illegitimacy really means.[6] A child born out of wedlock is stigmatized as illegitimate, but only because of laws made by men to navigate a fundamental problem for them, their own role in the paternity of children born to their sexual partners. Within legal marriage, regardless of whether or not a wife partnered with someone other than her husband, children are assumed to be the husband's in Japan (Article 772 of the Civil Code). But the question of paternity of children of unmarried women has troubled Japanese men and women for centuries, and it is of course also a problem in most other societies as well. Before DNA testing, men could only hope to control the sexuality of wives, but the rules of paternity and legitimacy allowed them to claim the offspring of women legally bound to them. In Japan, extramarital relations between married men and their unmarried concubines, mistresses, and so forth have been well documented and to some extent tolerated and naturalized, not only as inevitable but useful throughout history, and representations of these normative relationships can be found across historical studies as well as in fictional accounts from the Imperial Family on down.[7]

To understand the bias toward a male head of household that lingers in the Koseki Law and the Civil Code today we ought to consider the complex attitudes toward family, marriage, and nation developing during the late Tokugawa and early Meiji periods when these laws were coming into their modern form.

The Civil Code, finalized in 1898 some 27 years after the Koseki Law (1871), actually followed the Koseki Law and *koseki* apparatus, rather than prescribed it, but in its current form, the Koseki Law often points to the Civil Code for its validation.[8]

The worldview in late Tokugawa and early Meiji was based on an extensive hierarchical social ranking determined by occupational class, gender, and age. Although class hierarchy, based on occupation, was abolished in early Meiji, gender and age hierarchy were maintained and manifested in Koseki Law and the 1898 Civil Code. The five pillars of this hierarchical Meiji-era structure were:

1 registration of a family under a single address and single surname as the documented unit of society;
2 registration for multiple generations in one document;
3 registration in the name of a single individual household head, ideally, the eldest son in the lineage;
4 registration of all members of the household in terms of their relationship to the household head; and
5 inclusion of outsiders in a registration (usually marrying-in wives, but also, occasionally, marrying-in sons-in-law, or already married couples) to maintain patriline through reproduction.

Although the Koseki Law and the Civil Code have been updated since the Meiji era, these components continue in the current *koseki* system, spare one, the multiple generations on a single *koseki* and the notion of marrying in to a multi-generational patriline because in its current form the Koseki Law prescribes that upon notification of marriage, a new marital *koseki* shall be established, ending the appearance of patrilineality that showed more visibly on a multi-generational document.

1 Registration as a single unit

The registration of the family as a unit is fundamental to the authority vested in the head of the household by the state in Meiji. The historical developments in early Meiji situated the privilege of a male head of household squarely within the structure of the developing *koseki* system (the laws and practices as they were formulated during Meiji).

> Like the Western systems for recording and documenting individual civil status, the *koseki* was a tool to gain control through identification of the population. Yet, where the western civil status registration systems took the individual as their basic unit, the Meiji-era *koseki* compiled registers using the unit of the "*ko*," or, more precisely, by grasping clusters of individuals centered on a "household head" (*koshu*).
>
> (Mori 2014: 60)

Many subtle changes have been made to the Koseki Law since 1871, but registration as a family unit has not changed. This understanding of family is at the heart of the debates about separate surnames and identity in marriage, particularly for married women who are expected to enter or join the husband's family.

2 Registration through multiple generations

Keeping *koseki* through multiple generations by adding wives or sons-in-laws allowed a continuity of household through name, place, and identity over long periods of time, regardless of who actually parented the children and grandchildren. Meiji households tended to be patrilineal and patrilocal. Wives married into a family, a name, and a household and the expression *ie ni hairu* "entering the family" or *koseki in hairu* "entering the *koseki*" reflects the sense of coming in from outside. This practice meant that ideally an eligible son would become the head of household upon his father's death or retirement and stay in his natal household throughout his life. In this scenario his name would not change and his own son would have the same surname as his father, mother, and wife. His wife by contrast would leave her natal home (be crossed off her father's *koseki*) and become part of his household, adopting his name, his family's practices and norms (often down to the ways of making the morning meal and other household idiosyncracies) and become almost property of the family, her service as housekeeper, mother of an heir, cook, nurse to the elderly, and perhaps storekeeper, farmer, or helper in the occupation of the household. The expression "borrowed womb" was used extensively in the Meiji era to refer to a marrying-in wife who was seen as necessary for birthing an heir, but otherwise rather interchangeable with other females. She was borrowed from another family and her main asset was her ability to birth the heir. In the case of *mukoyōshi* (marrying-in son-in-law) a daughter might stay in her natal home and see one of her children become the future head of household, but it was still her own father's patriline and her husband might be considered more a son, than a son-in-law, to her parents.

Multi-generational households could demand duty and filial piety in the name of the household, de-emphasizing the wishes and needs of individuals. The multi-generation *koseki* is gone now, abolished in the postwar period, as one of the ways to curtail the legal status of the *ie seido* family system. But as a mental schema of family, a way of thinking about marriage and family, the multi-generation *koseki* seems alive in the ways that people think about and talk about their families and they ways they move in and out of other families. Because a father's full name and a mother's first name are always noted next to an individual's name on a *koseki*, the lineage is quite easy to trace even if multi-generation *koseki* have been abandoned in the postwar period.

3 Registration under an individual household head

Hierarchy within the household was a building block of the modern Meiji state. As the legal scholar Watanabe Yozo put it, "The political system after the Meiji

Restoration was not democratic but authoritarian and demanded that the people be uncritically obedient and docile" (Watanabe 1963: 369). Furthermore, the hierarchy and docility must be learned in the home, "The discipline a man undergoes within his family from infancy on forms his character and his morals" (Watanabe 1963: 369). Filial piety, a central feature of family attitudes and feelings, was linked to legal status and the role of the head of household in the Meiji law.

> Although filial piety was owed by the child to both parents, the exercise belonged in law as well as in fact to the father alone; only under exceptional circumstances, when the father was dead or otherwise incapacitated, did the mother exercise such power.
>
> (Watanabe 1963: 368)[9]

The head of household as established in the Meiji Civil Code of 1898 was based on the hierarchical structure modeled in the *koseki* document and the Koseki Law of 1871. The Meiji Civil Code specifically identifies the obedience due to the father in modern Japan. Beyond family order, "the government largely relied upon family discipline to bring up citizens who would submit to governmental authority" (Watanabe 1963: 369).

The household head in Meiji was the embodiment of authority. Vested with rights and responsibilities to maintain discipline, filial piety, and national duty, the head of household was granted his role as an extension of the state (Chapman 2014). The Meiji-era term for the head of household, *koshu*, continues to be used to express the kind of patriarchal authority granted fathers through law and social norms in the Meiji past and sometimes in contemporary society as some of my 40- and 50-year-old interviewees showed as they sprinkled the term into our conversations. In the postwar *koseki*, the household head position has been renamed from the patriarchal term of *koshu* to the benign-sounding *hittōsha*, first registrant, an apparently neutral term that is not only gender-neutral, but also cleansed of the Meiji insinuation of patriarchy. But despite the update, the *hittōsha*, like the *koshu*, still structures the *koseki* as a hierarchical document with only one person named on top and this person stands for the family through his surname. One of the telltale signs of the meaning of *hittōsha* is that even when he dies, a *hittōsha* continues as the head of household hittōsha a new register is created by the remaining family members, and often they never do that, because it is easier to stay within the male-headed household (Sakakibara 1992).[10] Clearly, the set of attitudes and feelings conjured in the old terminology *koshu* is not effaced or eradicated in the new term *hittōsha*. Many of my informants, including Sumie, used *koshu* and *hittōsha* interchangeably to signify male authority in the family and society. Others continue to use the word *koshu* in order to repudiate the position itself (Tanaka 1995). Thus, regardless of the term used to designate the head of a family (and the top of a *koseki*) a father's authority and responsibility to rule his home and family members can be seen on the document in use today. Many of the people I interviewed commented that since the *koseki* document cannot accommodate two adults as the top, sharing an equal status, such a family structure is difficult for people to imagine. As the *koseki* is now strucutured, only one person stands in for the whole family and only one person's name can be carried through for all family members. It is expected that for married couples the name will be that of the husband and in cases of a female-headed *koseki*, like Tanaka Sumiko's, it is understood that she must be an unmarried mother. Given that fewer than 4 percent of marital *koseki* are registered under a woman's surname, the linkage of male authority and patrilineal household structure is extremely widespread.

4 Registration of members in terms of their relationship to the household head

The *tsuzukigara*, the succession column discussed in Chapter 3, provides another site on the *koseki* for an investigation of underlying meaning of family structure and control. The information encoded in that space specifies the relationship of the child to his/her parents, as eldest son, eldest daughter, and so forth. This column mattered in Meiji Japan at a time when a preference for primogeniture was being legally established first on the *koseki* and later in the Civil Code. Single-child inheritance guaranteed the convention that property and land would not be divided upon the death of a head of household, and that other children or their spouses would not make claims on a household, allowing continuity for a farm or other kind of household business. This important rule was then firmly sutured to male authority with the expectation and preference stated in the Civil Code for not just single-child inheritance, but specifically, eldest-son inheritance, primogeniture. By specifying a preference for eldest son inheritance, care for and privilege of the eldest male child was established as a norm for the good of the family (*ie no tame*) that continued well into the late-twentieth century.[11]

5 Marriage for the continuity of the family line

Borrowed womb 借り腹 *karibara* is a term that was used to refer to women and concubines alike during the Meiji period and well into the middle of the twentieth century. Rather than gaining a wife, a man gained access to his progeny through a well-chosen woman who would simultaneously fill the roles of wife, daughter-in-law, mother, and worker in the household. A woman's labor was wildly valuable to a household, but it was for her "womb" or fertility that she was brought into a family and a *koseki*.[12] As mentioned previously, it was possible to adopt a whole range of outsiders for the purpose of family continuity for the *ie*. Article 970 of the Meiji Civil Code specified primogeniture as the preferred form of succession in the Japanese family but adoption of son-in-law or married couple was also possible. This flexibility to go beyond the patriline and the bloodline in order to carry on a family name and occupation, whether farming, trading, or some other family business, helped keep property within families and ultimately establish wealth during this period of great social transformation.[13] In the Meiji multi-generational *koseki*, such adoptions of successors allowed continuity through the generations. Continuity occurred at multiple levels, including at the level of a household (the physical house), at the level of the documentation of the household, the *koseki* – the document that represented the family and its members – and at the level of the ritual practices of care for household ancestors.[14] These levels of family practice were prescribed in law through the Koseki Law and the Civil Code and generated a sense of national duty and national reckoning, linking familial filial piety with Imperial obligation and respect. The *koseki* document then became a map of family relations and family obligations, structurally highlighting eldest males through the generations. The Meiji-era notion of family, *ie seido*, included not only the people who constituted it, but also the work they did as a unit, such as farming, shopkeeping, or production of goods and the identity (or family name) that defined them through generations.

Family continuity and Meiji sexual norms

Family continuity in Meiji depended on an eligible male successor and a reliable and fertile wife who could work with and for the family/household. The necessity of obtaining a successor or male heir provided a rationale for not only marriage but also concubinage. Extramarital relations played a role in the guarantee of an eligible heir in wealthy Meiji-era households as it had in earlier periods and as was modeled by the Imperial Family, a dynasty that required worthy male heirs (Shimazu 1994; Lebra 1993). In her study of the Japanese aristocracy, Lebra writes of this rationalization in the Imperial Family: "[i]deological justification for concubinage, which has continued into the modern period, derived from the mandate of the patrilineal succession of the *ie*" (Lebra 1993: 224–225). Continuity across generations provided a guiding principle in debates about the *koseki* law in the Meiji era also. To secure a good heir a man of means could increase his chances by having multiple mistresses or concubines. Though adoption was neither forbidden nor rejected as a possible option, concubines and their fertility provided a greater chance of a male having an heir of his own blood. Blood mattered, but was not essential for an heir as the Civil Code makes clear by allowing adoption of non-kin for succession. Concubines increased chances of a child born to the head of household, but needless to say, fertility was not the only incentive for obtaining a concubine.

Indeed, the rationalization of concubines as guarantors of male heirs obscured the fact that having a live-in mistress or concubine in addition to a wife was quite desirable. It is important to remember that marriages among the aristocracy, the samurai class and the wealthy during the late nineteenth century were generally arranged for the good of the family, not determined by an individual son or daughter's personal tastes or preferences. For the most part, parents made decisions on their children's behalf and even well-to-do young men probably had to abide by their parents' practical and expedient decisions. This pragmatic and relatively unromantic approach to Meiji-era marriage for the wealthy and elite may help explain the propensity for heads of households to engage outside or live-in mistresses during this period. In her fictitious portrayal of a Meiji government official's household during several decades of the Meiji period in *The Waiting Years* (*Onnazaka*), Enchi Fumiko explores the fraught dynamics of the lord Shirakawa Tomoyuki, his young, beautiful wife Tomo, of whom he quickly grows tired, his first concubine Suga, still a teenager upon her arrival to the household, and then a second concubine, Yumi. In Enchi's handling, the lord's ability to keep the women in the household at his beck and call emotionally and sexually was clearly motivated by goals beyond his progeny. Indeed, in the novel, Shirakawa already has an heir. Enchi's critical portrayal of a sexually voracious husband who oversteps even Meiji-era sexual norms brutally depicts the normalized concubine system of the time.

Concubines are importantly related to the study of the formation of the modern *koseki* in the Meiji period because they reveal understandings about marriage and the nature of the family at the time which still linger in the *koseki* today. In particular, marriage in Meiji was understood as an arrangement to guarantee lineal continuity and multiple "wives" could be rationalized in the name of continuity. Most of the children born to wives were expected to leave the household upon reaching adulthood so that only one child would carry on the male family line into the next generation. The eldest-son–maternal bond was structurally stronger than the husband–wife tie, leading to complex relationships between mothers-in-law and their daughters-in-law. More importantly, marrying-in wives were often only one of a number of types of sex partners for the male head of household. Men's sexual freedom and promiscuity hardly needed justification in early Meiji, because polygyny in the form of live-in concubines had been widespread during the Tokugawa period from 1603–1868, but as the inherent judgment of western eyes encroached on social and sexual norms in Japan, household succession, the need for a male heir, provided some cover (Lebra 1993; Fuess 2004).

Concubines had been common in ancient China and ancient Japan, though attitudes and practices varied from place to place and throughout history. In Japanese, the term *mekake* implied a kept sexual servant or mistress. While mistresses living outside a household were also common they were in a different category of social status from live-in concubines who had been placed with a family through some kind of economic or social arrangement. In the Imperial Household the use of concubines has a long history, as discussed by Sugiyama Lebra, and to some extent the institutionalization of concubinage in the Imperial Family set the standard for the rest of the society. In the Meiji period, as the emperor was restored to power and his family became a model for the nation, the practice of concubinage became standardized in modern law (Lebra 1993).

The functional explanation for the need for concubines should not mask the fact that men had a kind of sexual freedom that depended on women's lack of economic mobility and freedom. For the most part, young women did not choose to become concubines or lovers of married men in this period. Like the characters in Enchi's novel they were often sold to a wealthy man when economic times were hard (Fuess 2004; Shimazu 1994). Male privilege structured every fiber of the modernizing state and these privileges were also knit into the very structure of the *koseki* documents and system, starting with the status, reinforced through the Koseki Law, of a male head of household, *koshu*, who was vested with the authority by the state to make decisions on behalf of his family and to make entries or deletions to the household *koseki* (Mori 2014).[15]

The status of children born out of wedlock in Japan today is inextricably linked with the norms laid down in law during the Meiji era. The hegemony of men's sexual prerogative is evidenced in early Meiji laws including the 1873 edict on illegitimacy of children, the first legal instrument to specify children in relation to the marital status of their parents in Japan (Shimazu 1994). The Dajōkan Fukoku, no. 21, an 1873 edict of the young Meiji government, proclaimed, "[a]ll children who are born to women who are not [*saishō*, 妻妾] (a term referring to both legal wives and 'legal concubines') shall be treated as illegitimate and the mother will take custody of those children" (Shimazu 1994: 86).[16] This simple statement in this early Meiji period document reveals the categorization in law of three types of female sexual partners, legal wives, legal concubines, and other sexual partners (who presumably do not have any legal status). By declaring that the concept "illegitimate" would refer only to those children who are born to women who are neither legal wives nor legal concubines a third category of mothers of the illegitimate was specified. Embedded within the 1873 edict is the assumption (and the approbation of the assumption) that men will father children with women of various statuses; namely, wives, concubines, and other non-registered women, and that the fate and categorization of the children shall depend upon the status and registration of their mothers.

In her discussion of the 1873 decree, the historian Shimazu Yoshiko examines the ways in which the leaders of the new Meiji state concerned themselves with this categorization of children, making the distinction between wives who were registered (revealing a kind of legalized polygyny in place at the time since there could be more than one *saishō*) and sexual partners who were not registered. *Saishō*-妻妾 referred to all "legal wives." The character 妻 *sai* or *tsuma* specifies wife and the character 妾 *shō or mekake* means concubine. The compound term connotes legitimate wives and concubines, suggesting the normalcy (and indeed legitimacy) of keeping concubines in contrast to non-registered (non-legitimate) sexual partners and their children which would in the 1873 legal rendering be considered "illegitimate."[17]

In other words, the edict regarding the status of children born out of wedlock took for granted that concubines could be registered "like legal wives" during this period. Thus Meiji leaders at least initially sought to normalize and legitimate a type of polygyny in modern family law by protecting concubines as "legitimate" mothers who could potentially give birth to an heir or head of household. But as Shimazu shows, the protection of concubines and their children was short-lived. Concubinage remained a protected family status for only 12 years in the Meiji period from 1870 to 1882. From 1882 forward, concubines lost their legal standing in the family, and a child born to anyone other than a legal wife (and from thenceforth there could be only one) became designated as born out of wedlock or illegitimate.[18]

Shimazu points out that during the 12 years (1870–1882) when concubinage was legally protected, a child born to a registered concubine, upon the decision of the male head of household, could be designated as his heir, not only legitimate, but the one child who would head the family in the next generation in the system of primogeniture in place at the time (Shimazu 1994). In other words, legal status enjoyed by concubines at the time allowed legitimacy for their children in the case that the male head of household chose to grant it, a further demonstration of the legal privilege embedded in the head of household position. Such a system made it possible for a man to guarantee his succession in the next generation, and purportedly, this was the logic behind the rule (Shimazu 1994).[19] The head of household's power as arbiter of the fates of others, including both registered concubine/wives and unregistered female sexual partners, was supreme. As a result he had the power to determine if a women would become mother of an heir enjoying high status in his household or mother of a "bastard" or "illegitimate child" (a new category known as *shiseishi*, privately born child).

The pejorative term for out-of-wedlock birth, *shiseishi* or privately born child, was used to signify that such a child would neither be placed in the male lineage nor in the male-headed *koseki*. Children in the father's *koseki* were by default considered publicly born children (*koseishi*) assumed to be the children of the head of household and his legal wife. A measure such as the 1873 edict permitting the son of a non-wife (but legal concubine) to sustain a family line was acceptable. But clearly, in the many cases when a non-marital child would not be designated as an heir, the Meiji leaders believed it was necessary to clarify their status and specify their inferior position. Prior to 1873, the identity of children born out of wedlock had never been characterized in legal terms (Shimazu 1994). Indeed, during the prior Tokugawa period treatment of out-of-wedlock births varied across different regions, including the common practice of registering such children in a relative's household as a son or daughter. The designation as "illegitimate" was a Meiji invention coextensive with the *koseki* document (Shimazu 1994).[20]

As activists fighting to eliminate the notation of illegitimacy on the *koseki* in recent decades have argued, the stigma of illegitimacy persists in Japan today. Article 790 of the current Civil Code (2) links illegitimacy with the *koseki* most directly, stating, "*chakushutsu de nai shi wa, haha no shi o sho suru*" ("A child born out of wedlock shall take the surname of the mother"). Taking one's mother's surname means being registered in one's mother's *koseki*. This part of the Civil Code essentially excludes a non-marital child from its father's *koseki*. The surname is used as the index that guarantees different *koseki* for the mother and father of the child when the parents are not legally married to each other.[21] In such a case, the father may be unmarried, married to another woman, still in his parents' *koseki* or in a *koseki* alone, but the child born to his non-marital sexual partner may not be registered in his *koseki* because of the distinction of legitimate and illegitimate in Civil Code 790 that originated in the Meiji period. Exclusion of the out-of-wedlock child becomes a powerful incentive for abortion or shotgun marriage when possible.[22] Activists choosing to let their children be designated as born out of wedlock because they want to keep separate surnames and resist legal marriage upset the intended logic of the Civil Code rule. Government authorities and conservative lawmakers would not expect any citizen, even critics of Japanese patriarchy, to risk the well-being of their children over the principles of equality and social justice.

Inheritance rules and acknowledgment of a child

Beyond overt discrimination based on the *koseki* there are financial and symbolic aspects of the treatment of children born out of wedlock that have been challenged by activists in the court. One of the most overt legal signs of the unequal treatment of children born out of wedlock can be found in Article 900 of the Civil Code, which until several years ago stated that all children shall be eligible for equal portions of inheritance, except children born out of wedlock who are eligible for one-half of a legitimate child's portion. When I interviewed *koseki* lawyer, Sakakibara Fujiko, in the spring of 2013 she said that this particular Civil Code provision is a key component of inequality of children born out of wedlock in the Civil Code and a sore point for many elderly people whose biological father provided some inheritance, but not an inheritance equal to his legitimate children. (Of course, in a culture that uses wills to settle such matters, the idea that the illegitimate offspring have rights to equal inheritance is fascinating.) Our conversation was just months before the Supreme Court handed down its decision on several cases pertaining to Article 900, deciding in favor of the plaintiffs and ending the very bitter "one-half rule."[23]

Understandings of inheritance in Japan transformed after the end of World War II when the norm of succession, primogeniture, was altered through a Civil Code change by the Occupation members working on family law and Civil Code. In the newly mandated more egalitarian division of property by law, spouses were entitled to one-third of a head of household's estate and children were entitled to equal parts of the remaining two-thirds, except illegitimate children, who were specified as eligible for one-half of a legitimate child's portion in Article 900. In order for children born out of wedlock to receive any portion of their biological or presumed biological father's estate they had to have proof of a kin relationship. This proof resulted from a process of identification of a father on legal documents known as acknowledgment or recognition, 認知 *ninchi* in Japanese.[24] In Japan, in order to create this legal relationship, a father is required to submit paperwork called an Acknowledgment Notification (*Ninchi todoke*) form, specifying his paternity. This notion of acknowledgment entered Japanese law just at the time when concubines, *koseki* registration, and other family matters were being codified in 1873, making it possible for a man to acknowledge his paternity on a child's *koseki* (headed by its unwed mother) and establishing a relationship between a child and its father.

The Meiji status of "acknowledgment," which designated a child as a *shoshi*, is now referred to through the process of *ninchi*.[25] The most common type of *ninchi* today involves a father who acknowledges a child born to his mistress or lover. In such a case, the unmarried woman will register her child born out of wedlock in her *koseki* with information about the child's father, and details about the location of his *koseki* based on his completion of an Acknowledgment Notification. In the father's marital *koseki* with his legal wife, there will be a note that he has acknowledged a child registered in another person's *koseki*. Obviously, such declarations are not good for a marriage and we can assume that many children born out of wedlock are never acknowledged because of the difficulty of placing the information about them in the father's marital *koseki*. Nonetheless, there is a sense of duty to a mistress and her child and the method is not uncommon. The information on the Acknowledgment Notification form and *koseki* links the father to his child and means that upon the death of the father the child is entitled to some portion of inheritance. *Ninchi* also provides the information about a father to a child who is possibly being raised by a single mother and may have no contact with his father. There are many different scenarios for this kind of relationship. But it is common for a child born out of wedlock and registered that way to have minimal contact with a father when he is legally married to another woman. Indeed, it is not uncommon to keep a child from knowing who the father is, in part by keeping the child from seeing his or her *koseki* until marriage or at least reaching the age of majority, 18 years of age, because the documents are generally handled by adults and there are often family secrets that are not shared with children.[26] Although there seems to be little research on this issue, my conversations with interviewees and acquaintances suggest that we can assume that many children do not know that they have been acknowledged by a man married to a woman other than their mother. Thus, it is possible that a child is unaware of the father's name and whereabouts during childhood. Importantly, acknowledgment cannot make a child born out of wedlock "legitimate" or *chakushutsu ko*. While a man may treat both his wife and mistress "like wives" the law requires that only those children born to his legal wife shall be granted the legal status of "legitimacy."

During the Meiji period the treatment of the acknowledged child underwent several changes. But once legal concubinage ended, any children who were born to a concubine or other non-wife were designated as *shiseishi* and that status increasingly became a blemish on their identity and *koseki*. The 1873 edict had required that any non-marital children be registered in the mother's *koseki* and the location of the child's registration became the proof of its status. Although a father had power to acknowledge and register a child as his own for succession purposes until 1886, a revision to the Koseki Law that year made it clear that the registration for a non-marital child had to be in its mother's *koseki* (Shimazu 1994). From that time on, the location of a child's registration, either in a single mother's or in a married couple's *koseki*, became the determinant factor of the child's status. This logic continues today as the location of a child's register in a marital *koseki* with its mother and father or in the *koseki* of its single mother reveals the child's legitimacy status.

To clarify the role of modernization (and indeed westernization) on the status of children in the Civil Code, we should compare the Tokugawa norms to the

Meiji norms. During the Tokugawa period children born out of wedlock were absorbed into families without demarcation of legitimacy status (Shimazu 1994; Toshitani 1994). But in Meiji the change of responsibility for a child's registration from the father's *koseki* to the mother's by law suggests a shift in thinking about social and sexual norms. In early Meiji, men had easily claimed more than one woman as a wife. When the law changed, outlawing a second wife, it essentially required monogamous marriage and began to treat not only second wives or mistresses, but also their children, as outsiders, specifically outsiders to the family, a condition reflected in the *koseki*.

The male prerogative to claim an illegitimate child as needed in Meiji compounded a non-wife's vulnerability; she might lose her child if the father chose to claim her child as his own to be raised in his household by his legal wife or she and her child could be ignored, left with no legal rights. At the same time, a legal wife might be asked to raise her husband's lover's child as her own. Attitudes and understandings of sexuality, particularly male sexuality and concubinage, in light of western influence, were being reshaped for a modern, increasingly westernized era, just as the *koseki* and family law were taking shape in the Meiji era. These responses to rapid modernization shifted the normative family in important ways, first legitimizing concubines and then rendering them and their children outsiders, and reveal the low regard with which women as mothers of children, whether legitimate or not, were held.

Male privilege and social inequality

Male perspective (as Sumie called it), or male privilege as we might call it, is at the heart of *koseki* structure and logic. The concept of privilege has been elaborated in American feminist analysis to explicate the often unspoken and taken-for-granted manifestations of social inequality in US society. Feminist scholars, working to further specify the nature of power dynamics, have elaborated the complexity of interlocking forms of oppression. Patricia Hill Collins, American sociologist, suggests understanding power dynamics as a *matrix of domination*, a construct that helps elucidate the interlocking factors of gender, race, and class that situate people differently within a shared social structure shot through with power dynamics that result in inequality and social injustice.[27] In Collins's framework no one is outside of power relationships. Rather, membership within certain groups (heterosexual, middle-class, white, female, or Latino, middle-class, gay, for example) situates people at times in positions of relative privilege and power and at other times in positions of subordination. According to this framework for thinking about power, most people will some of the time find themselves subordinated to those in more dominant positions within particular contexts, but at other times find themselves in a relatively dominant position vis-à-vis others. Collins emphasizes that people are not in static positions of dominance or subjugation, but rather in positions of relative power based on the totality of their identities. A factor like whiteness, in a society like the United States, that provides myriad advantages for people of European descent, may have different effects if a person with white privilege also happens to be transgender, poor. or gay. In other words, no single axis of identity will necessarily prevail in a given context or life. As situations and contexts vary, so too will the meanings of one's identity and one's ability to move freely in society.

The concept of privilege provides another tool through which to understand social hierarchy and ongoing inequality. The sociologist, Peggy McIntosh, in an attempt to isolate a pernicious, and to some people, elusive aspect of racist and sexist social structures, in her now classic essay of the 1980s, shows some of the ways in which white people benefit from not only a broad and extensive assortment of social structures that favor them, but also assumptions about who they are and what they do. McIntosh's work helped spark a critical, multi-decade discussion of the role of privilege within US society, highlighting privilege's nearly invisible nature to those who benefit from it most. She writes that again and again whites, men, and other privilege holders tend to ignore or deny the many advantages they have in US society. In her essay, McIntosh describes a series of daily life norms that are taken for granted by most whites, that include being assumed to be innocent, good, and reasonable (at the bank, in a traffic stop, or at a child's school, for instance). Like white privilege, Mclintosh points out, male privilege too, works and survives in part through its invisibility to those who have it. Privilege and unearned advantages operate like an invisible knapsack filled with "special provisions, assurances, tools, maps, guides, code books, passports, visas" and so forth that make movement through the world easier for those who are part of a privileged group (McIntosh 1988). Other scholars examining male privilege have shown the many ways in which society's norms are often based on an assumed male citizen and what fits his life course, body type, attitudes, and tastes. In medical research in the United States, for instance, we can see that a default "average" patient on whom research is based is often male, white, young, and healthy. Of course, there may be good reasons to choose young individuals for research trials, but the extent to which women and people of color have been overlooked in medical research and drug trials is just a further example of the biases built into the structure of our society and the concomitant assumptions about who an average citizen is.

Legacy of Meiji-era male privilege in Japan

Needless to say, privilege is at work in the norms and structures of Japanese society today. In the home, workplace, the public and private spheres, we can find many examples of male privilege. The Japanese Diet is still made up predominantly of men, with women occupying just 16 percent of the seats in the Upper House and 9.5 percent of the seats in the larger, 475-member, Lower House of the Diet. Corporate Boards in Japan are dominated by men and even the percentage of women as CEOs of major corporations continues to be extremely low, less than 10 percent. In addition to this, women do an overwhelming amount of care work for the young, the sick, and the elderly in Japan. Much of this work is unpaid and done in the name of family (Kawano *et al.*

2014; Long 2014). As a component of family membership, care activities are rarely understood by government or public institutions as work. The effects of women's unpaid labor on their lifetime earnings, career opportunities, and creative contributions to society are stifling. In care work we see women overwhelmingly paying the price of gender expectations that contradict legal equality in the Constitution and of course this is not only a problem in Japan, but Japan's problem is extensive because of a compounding of many factors, including rising average longevity, women's increasing participation in the paid labor force over several generations, expectations that daughters and daughters-in-law will take charge of elder care, the overwork of many full-time workers, both male and female, which aggravates gender norms and roles in the household, and more. Although the care of dependents is the concern of the entire society, institutions in Japan have relegated the responsibility for that work to women as members of the family, fulfilling expectations as daughters, daughters-in-law, mothers, and wives.[28] Through familial duty, a sense of obligation and love, not to mention an overwhelming sense that if they do not do it no one else will, they pay a price that neither the state nor private sector is willing to pay.

It is difficult to imagine the tables being turned in Japan. Can we even imagine a Japan where men assume the unpaid labor of caring for children, the sick, and the elderly, while women move independently choosing their own path and securing jobs that suit their interests and abilities in society?[29] In Gramsci's notion of hegemony those who are oppressed share the values (ideology) of those who oppress them and perpetuate their own subordination through shared values. According to him, ideas and practices which benefit the bourgeoisie are taken on by the working class as their own and the status quo is maintained rather than challenged. We can see a similar phenomenon with gender roles in contemporary Japan. The assumption that childcare and care for the elderly are women's work is so widespread and taken for granted throughout society that we can say it is hegemonic in Gramsci's meaning. Not only is women's responsibility for the care of all weaker and dependent people naturalized and taken for granted, but it is rarely questioned at a fundamental level. Truly questioning and addressing whose responsibility this substantial care work is would upset the whole Japanese social structure and economy. To the extent that women are trapped in this system and discouraged from questioning it, it is hegemonic. The ways in which men benefit from this hegemony should be obvious to anyone who has ever changed an elderly relative's diapers, gone to work only to come home to cook for family members, bath and care for children before cleaning and putting children to bed, or to anyone who has been told that her commitment to her job is inadequate or her unwillingness (or inability) to work long hours shows her inadequacy for a job with more responsibility or higher pay. Without extensive cooperation from an unpaid partner or a paid surrogate, it is very difficult to cover all the care work that needs to be completed in contemporary society, whether in Japan or elsewhere. Male privilege gained a stronghold in Family Law during the Meiji period, and to the extent that the privilege has been hegemonic and relatively invisible it has been nourished into new and myriad forms in contemporary society.

The Meiji-era understanding of unmarried mothers and its legacy for unmarried mothers today

Activists working to change attitudes about children born out of wedlock in Japan like Tanaka Sumiko and Tomizawa Yoshiko are in relationships *distinctly* unlike those of the women whose children were deemed *shiseiji* or illegitimate in the Meiji past. By choice, today's activists have stayed out of legal marriage. Despite being unmarried, they are in long-term, monogamous relationships with the fathers of their children. Through the decision to keep their own names and stay unmarried they have found their children perceived, or treated, like the out-castes to a family of children born to concubines or mistresses. This mistreatment and indeed mis-recognition of who these children are reflects past biases. The time of concubines and kept mistresses has mostly come to an end in the twenty-first century. Compared to the Meiji period and even the early postwar period, women have many more options of how to live. While the notion of filial piety continues to be a powerful motivator to marriage and career decisions, in a wealthier, modern Japan, young women today are highly unlikely to be sold or traded by their parents into the households of wealthy men as was so common in the Meiji period.

But the point here is not that the system of concubinage ended and that women in Japan today have more sexual autonomy and more say in the types of lives they will lead, but rather that unmarried women of the twenty-first century, especially those with children, continue to be labeled in categories that were created at a time when concubinage and kept mistresses were the norm, a time when "good women" (read married) were made distinct from "bad women" (read unmarried, sexually active ones). The powerful and pernicious whore/wife dichotomy which condemns unmarried women and praises the purity of wives continues today, despite tremendous transformations in attitudes about the institution of marriage and female sexuality. Japanese feminists exasperated by the sexual double standard in Japan have tried to expose this fundamental and deep logic in their work for decades (Tanaka 1975).[30]

To return to the problems in the *koseki* today we can start with the designation of "*chakushutsu de nai*" on the Birth Notification form. This designation is a reminder of social attitudes about children that fathers have not claimed. A false equivalence between a kept woman whose child was shunned as *shiseishi* (or *shoshi* if acknowledged) in the nineteenth century and an activist like Tanaka Sumiko who has a child born out of wedlock by choice, leads to mis-recognition or mis-categorization on the *koseki* and the Birth Notification form. As Tanaka has argued, her child was labeled as worth "less than a whole child," positioned as inferior to other children on the *koseki* and *jūminhyō* because of a system developed in the nineteenth century when multiple wives and multiple statuses of children were the norm (Tanaka 2004a).[31] The legal protections of a married man and the vulnerabilities of an unmarried mother and her child are at the heart of male privilege in the *koseki* and in Japan more generally.

Thus, to speak of the discrimination against children born out of wedlock, for the activists who reject the constraints of the family system, is to engage with a larger set of issues about not only historical family norms, but also attitudes toward male sexuality and the failure of male lawmakers who dominate the Diet to revise the laws to reflect changing social and sexual mores. As my informants so eloquently argued, the main site of the antiquarian Meiji notions of family are felt and experienced through encounters with the *koseki* document, which is backed by the hidebound Koseki Law and Civil Code.

The early laws identifying and specifying different kinds of mothers and different kinds of children were written at a time when the only reason a woman was not married to the father of her child was because he was already married to someone else and she was in a position to be a mistress, concubine, or lover, not a legal wife. The position of such women, often a result of socio-economic conditions, was to be marginal and secondary to a legal wife. Thus, the 12-year period of legal concubinage allowed a significant protection for registered concubines and even greater chance of support and status for the offspring. During this same period, primogeniture was becoming more and more consolidated and normative so that by 1898 it was the central pillar of the Civil Code and understandings of the family. In this sense, systematic and legal primogeniture went hand in hand with multiple "wives."

The fight for equality in inheritance

At the same time that activists like Tanaka and Fukukita worked to eradicate the discriminatory notations of children born out of wedlock on the *koseki*, others were working to end the inheritance law that designated a "one-half portion" to children born out of wedlock by suing the government for discrimination against children born out of wedlock. Article 900 of the Civil Code specified that all children should get equal portions of inheritance, except children born out of wedlock who would be entitled to one-half of the portion of a legitimate child.[32] There have been numerous lawsuits demanding that all children should be treated equally, and that children born out of wedlock are entitled to equal inheritance. While not a main focus of most of the people I interviewed, the issue of unequal inheritance was often mentioned by informants including Tanaka and Tomizawa. In reality, my informants' children would not suffer from the rule of half inheritance because these parents had no other children and the issue of one-half of the children's portion would be moot. But plaintiffs in several cases that were eventually heard by the Supreme Court in 2012–2013 were different. Now adults and in some cases elderly, they had been raised by single mothers and their "one-half" status had been made clear when their fathers died and they were notified of their father's paternity and inheritance.

If Japan were a society where legal wills and last testaments were widely used, perhaps the gravity of the one-half heritance clause in Article 900 would not have garnered such importance. With a will, a father who wanted to protect the child of a mistress could provide for the child more directly. However, since the will is uncommon, the *koseki* and the Civil Code function as de facto wills.[33] The Supreme Court decision of September 2013 granting equal inheritance for all children was heralded as a major step in eradicating discrimination against children born out of wedlock. The new rules which have been incorporated into law in the Diet in the past several years might be cause for some men to withhold their "acknowledgment" of a child born out of wedlock, knowing that it will impact the inheritance of any legitimate children. However, I have not heard of any such cases to date.

Conclusion

In this chapter concubinage, paternity, and patriline have been examined within a discussion of male privilege and gender hierarchy that developed through modern laws and social structures that were codified during the Meiji period while primarily privileged, educated men built the legal and economic infrastructure for a new society. When only certain members of a society form its Constitution, laws, and institutions it is no surprise that their perspective skews towards their own interests (Pateman 1988). The status of women in the family and in society at large was being determined for the benefit of the modern capitalist system in Meiji Japan, increasingly entangled with a global world order. In the codified family system established through new laws beginning with the 1871 Koseki Law, women were expected to move from natal register to marital register, divorce occurred at the convenience of the male head of household, concubines were welcome in the home to serve the privileged male head of household, and so forth. The battle being fought by feminists in Japan today is still tangled up with the powerful benefits laid down in the Koseki Law and Civil Code during Meiji. For although the Allied Occupation, with the help of the likes of Beate Sirota Gordon, believed that they were democratizing the Japanese family system and indeed eradicating the *ie seido* with Articles 14 and 24, they left not only the Koseki law and *koseki* document securely in place, they also left the matter of male sexual prerogative lodged within the columns of the *koseki*. The *koseki* continues to be structured by a man's relationship to his spouse and children. The naming of eldest son, eldest daughter, or adopted daughter, as in the case of Sumie Oshino, on the *tsuzukigara* reflects children's relationships to a head of household, a position held overwhelmingly by fathers, not mothers, which has been discussed at length in the previous chapters. While the *koseki* is called a household register or a family register, this neutral name masks the fact of men's privilege, the male viewpoint that orders the *koseki* document and *koseki* system.

Notes

1 Saeko helped me with with a wide range of tasks related to my research, including deciphering the legally dense rules specified in the *Koseki Jitsumu Tebiki* (Koseki Manual), drafting and revising my survey, and sometimes introducing me to people I would eventually interview.

2 See Buckley (1997) for an interview with Ochiai and a translation of a chapter of her novel, *The Rape*. In addition to the section on Ochiai, Buckley provides biographical background, interview transcripts, and translations of the works of nine other well-known feminists, who made their mark in the 1980s and 1990s and continue to shape Japanese feminism. *Broken Silence: Voices of Japanese Feminism* is an invaluable contribution to understandings of key feminist tropes and tensions in Japan.

3 Sumie also said that her mother would noisily climb the stairs to call her daughter to dinner so that her stepfather always had time to get off of her before her mother walked into the room.

4 In order to become adopted, Mr. Nomura would have had to formally adopt her, filing an Adoption Notification form at the local government office. Because the couple married and he was adopting his new wife's biological child, the paperwork would not have been complicated, but nonetheless, a child who had been registered as an eldest daughter on the *tsuzukigara* (birth order column) of her first *koseki*, was now moved into a new *koseki* with a new last name as an adopted child. The notation in the *tsuzukigara*, rather than daughter or eldest daughter, was *yōshi*, adopted child.

5 See White (2014) for further discussion of Satō's discussion of the gap between Constitutional ideals and on the ground koseki practices.

6 A *koan* is a Buddhist mental meditation puzzle, such as, what is the sound of one hand clapping?

7 *The Tale of Genji* (Murasaki, in translation, Seidensticker 1976) provides a good early representation of sexual mores that show polygyny as common among the aristocracy in the tenth and eleventh centuries.

8 See Chapman and Krogness (2014) for further discussion about the historical ordering of Koseki Law and Civil Code drafting in Meiji.

9 Meiji Civil Code, Article 877, states that "A child is subject to the parental power of his father belonging to the same house."

10 I was told on more than one occasion that it is easier to keep a deceased husband as a household head for a variety of tax and real estate reasons.

11 It seems fair to say that any reader familiar with Japanese culture will have heard stories of second, third and fourth children, regardless of gender – even though it was often most brutally felt by girls – overlooked or disregarded because of the special treatment and extra provisions kept for an eldest son.

12 Trial marriage during Meiji was one way to make sure a "good womb" had been acquired. See Fuess (2004).

13 A daughter whose husband became the new household head was usually in a relatively strong position because she would live in her natal home with her parents. Her family name stayed the same and her husband was adopted as a son in a practice known as *mukoyōshi kekkon*.

14 Book Five of the Meiji Civil Code specifies that land, property, and ritual instruments, such as the family altar for memorialization of the ancestors, were to be included in succession.

15 Although the Meiji-era *ie seido* is often understood as "traditional," as many historians and critics of Japanese family have noted, the patriarchal and patrilineal *ie* system that was codified in law during Meiji was a modern invention heavily influenced by values and norms of the samurai minority that favored single-child inheritance through the patriline (see Ueno 1987; Nishinomiya 2014).

16 The Dajōkan Fukoku, no. 21 (a decree by the Dajōkan, the predecessor of the Cabinet), of January 18, 1873.

17 See Fuess (2004) for a thorough discussion of the flexibility of modern marriage, particularly at its inception during this period.

18 The meaning of illegitimacy was defined through the designation of the mother and the mother's relationship to the father. It provided protection of men, so that they would not be held responsible for all the offspring of their sexual partners.

19 Also, see Lebra (1993) for a discussion of concubines in the Imperial Family where the need for a male heir rationalized a long tradition of kept concubines.

20 The *Zenkoku minji kanreiruishū* that catalogued customs at the end of the shogunate, stated that such children might be adopted by relatives, registered in either family's register, given to a childless family as an heir, or designated as a sibling of the father. However, "It is normal that no distinction is made between the legitimate and illegitimate child" (Shimazu 2014: 97).

21 If legally married, they have one surname (750) and one *koseki* (de facto 750). By definition (except in the ironic case of a man and his lover happening to have the same surname (such as White and White)) the surnames of the child's biological parents will be different because they are unmarried.

22 See Hertog (2009) for further discussion of the pressure to marry or abort to avoid giving birth to children out of wedlock in the twenty-first century.

23 At a meeting run by Tanaka Sumiko about ongoing activities to end discrimination against children born out of wedlock, activist Kodama Ritsuko, an elderly woman, gave powerful testimony about her situation as a child born out of wedlock. She had been raised by her single mother and had tried to acquire inheritance from a wealthy father upon his death. "I am not one half a person" she said. "I should not have received one half of other children's inheritance. The fact that my father could not marry my mother was not my fault."

24 At the same time that the category "illegitimate" came into law in early Meiji, the concept of "acknowledgment," an idea found in Germany, England, and France at the time, was introduced in Japan (Shimazu 1994).

25 There are several different types of *ninchi* possible today. The most common is the type for a child born to a non-wife. Chapter III of Book IV of the current Civil Code, minorly altered since it was originally drafted in 1898, delineates the Parent and Child relationship, in Japanese, simply Oyako 親子. In Section 1, Articles 772 through Article 791 deal with legitimacy and acknowledgment. Articles 792–817 concern adoption. These two areas of the Civil Code are keenly important. They reflect the concerns of those early Meiji years when the Koseki Law was in place and the rules of the Civil Code were being formulated.

26 The idea that the *koseki* is protected even within the family because of the personal and sometimes unsavory information it contains, in and of itself, is worthy of further consideration.

27 In early iterations of the matrix of domination Collins does not include sexuality or other spheres of identity and oppression, however, in later versions she does insert disability and sexuality as key axes of power and oppression.

28 For a fictional critique of the overwhelming expectations on women in the home regardless of their contribution to the family budget through part- or full-time work set in the 1990s see *OUT* by Kirino Natsuo (2003 [1997]).

29 There has been a trend of a small, but significant, number of young men dropping out of the workforce to care for their small children while their wives work outside the home and provide financially for their families.

30 Tanaka Mitsu clearly articulated the sexual double-bind and concomitant whore/wife dichotomy in her early second-wave feminist treatise, "Toilets No More" "*Benjo kara no kaiho*!" referencing the way men treated women in the left movement as nothing more than places to release their sexual energy, like a toilet. See Tanaka (1995) for a good overview in English of key moments in the second-wave movement in Japan.

31 Here Tanaka-san refers to the one-half inheritance rule of the Civil Code that was found to be unconstitutional in 2013.

32 When the Civil Code was re-written in the early postwar period in 1947 to reflect what was considered a more democratic family structure by the Allied Occupation, inheritance shared equally among all legitimate children became the norm and was specified in law. After decades of formal and legal primogeniture, the change in the

early postwar period guaranteeing inheritance to a surviving spouse and children was profound. Still, as I have heard from friends, informants, and students, informal one-child inheritance is quite common still, because of a desire to keep a house, property, or business intact.

33 I spoke with several informants about wills. One person told me that when her father died, the children and the children's children were notified of their interest in the estate. She said the children and grandchildren feared that someone unknown to them might approach the family as a child born out of wedlock. The question of the *koseki*'s use for inheritance and the infrequency of the use of wills is extremely interesting and should be explored further.

5 Beyond the scope of the *koseki*

Families out of bounds

Despite the elaborate governmental bureaucracy that supports the accurate implementation of the Koseki Law and Civil Code today, a variety of familial misrepresentations, some that have developed as loopholes exploited by activists and others that result unintentionally from the highly structured and heteronormative Civil Code and *koseki* system proliferate today. These misrepresentations include the absence of foreign spouses from the spousal column on the *koseki* (because they are not Japanese citizens), the appearance of parent–child relationships in the case of same-sex partner adoptions (a strategy for "marriage" where legal marriage is not possible), the appearance of being single when one is in a long-term common-law marriage, the appearance of being divorced living as married, the non-registration of children to avoid discriminatory designations of "illegitimacy," and the designation of "illegitimacy" for children who don't fit any definition of "born out-of-wedlock" as discussed in Chapter 4. These inaccurate familial representations expose the underlying fissures in a system developed within nineteenth-century understandings of family that inadequately account for the kinds of kin relations that proliferate in Japan today. Some result from intentional workarounds to avoid or subvert patrilineal norms while others result from inherent contradictions in a system designed to simultaneously document citizenship and family membership.[1]

When the material form of the *koseki* fits the configuration of one's family, the precise rules of registration may be easily accepted (without awareness of the prescriptive and limiting nature of categorical options), but for people whose family does not fit the hegemonic norm, the tight categories and rules can wreak havoc. For a couple I interviewed in the summer of 2014, two women registered in a civil union in Great Britain, and at the time living and working in the Kansai area of Japan, the *koseki* not only did not fit their family, their kin relations were beyond the *koseki*'s scope. Indeed, as a document of their kin relations, the *koseki* was woefully inadequate.

Mikiko Ishihara introduced me to Kaoru through email exchanges because of the situation Kaoru and her partner were in over the citizenship of their yet-to-be-born child. Mikiko had mentioned the paternity and citizenship questions that the couple faced. Japanese citizenship is determined by *jus sanguinis* (rule of blood or genealogy) and the couple's *koseki* issues were complicated by the fact that Japanese marriage rules do not recognize same-sex marriage.[2] In Great Britain, where Kaoru and her partner Lil had formerly lived, they had registered their relationship as a Civil Union, and thus considered themselves married, despite their non-legal marital status in Japan. Although I didn't have many more days left while in Japan for a conference in July 2014, I decided to take the train south to meet Kaoru and Lil and hear about their situation.

Kaoru met me at the exit of a small station not far from Kobe. Our first greetings were easy. She seemed familiar even though we had only corresponded through email. Like so many other feminists I know in Japan, Kaoru was warm, friendly, and generously willing to talk with me about her family and the *koseki* system. Her partner Lil, eight-months pregnant at the time, showed up a few minutes later and we proceeded to a nearby restaurant where we talked during lunch. I was eager to hear about the couple's efforts to gain citizenship for their child, knowing there were limited ways this could be accomplished. We talked about feminism, our connections in Japan, and our shared friendship with Mikiko. Kaoru told me that she has been critical of the Japanese family system (and the *koseki*) for a long time. "From a young age, I questioned the whole idea of marriage. Still, when I was young I started living with a guy who was still a student and we decided to get married. We had two children together." The financial and social conventions that made marriage in Japan heteronormative and pervasive (nearly compulsory) during much of the twentieth century provide the backdrop for Kaoru's decision to get married as a young woman. She says she was opposed to marriage, but slipped into it easily with financial enticements from her employer. "Since I had benefits from my company as a married person, it was economical to get married. It amounted to about one million yen a year" (1 million yen = $9,100/year in 2017, but approximately $25,000/year in the 1980s).

Kaoru told me that she had registered for divorce from her husband in order to resume use of her surname some years after they married, using paper divorce, but staying together with her husband to raise their children in the 1990s. Years later they decided to actually end the marriage and the divorce became real. Since they had already divorced on paper there was no need to file any additional paperwork at the local government office at the time, revealing some of the implications of paper divorce. Kaoru subsequently went to England where she met Lil and eventually married her in a Civil Union before returning to Japan for employment opportunities. Kaoru is a professor and has published in Japanese and English on a range of feminist issues. When they arrived in Japan they realized that their British marital status would have no legal significance in Japan. In the absence of a marital *koseki*, the couple can neither show a legal relationship nor prove that they are recognized as married in another country. For many same-sex couples in Japan this is a familiar problem. In a case like Kaoru's, however, where her partner is the biological mother and the non-Japanese member of the couple, the stakes are particularly high with myriad long-term consequences for the child born to a semi-married (according to British law)/ unmarried (according to Japanese law) foreign woman. And it was precisely the citizenship status of the yet-to-be-born child that was the concern for them when we talked in the summer of 2014.

Kaoru currently heads her own *koseki* since she is a Japanese citizen and divorced from her husband. Lil, on the other hand, is not named in any *koseki* in part because she cannot be named as spouse (wife) to her female partner and she is not Japanese. In other words, she has no place on any *koseki*. In order to have a child together the couple sought a sperm donor, one who would be willing to provide proof of Japanese paternity in order for the child to qualify for Japanese citizenship. Finding the right donor wasn't easy, but after discussing their situation with several candidates they found someone who understood that they did not need a third parent or family member, just a donor willing to provide acknowledgment through the process of filing an Acknowledgement Notification form (*ninchi todoke*) in order to register the child as a citizen with a *koseki*.[3] Since Lil, the birth mother, cannot show a legal marriage in Japan, the baby's citizenship would default to British unless the biological father was willing to acknowledge it as his own. Over lunch we discussed the search for their donor, a man willing to provide, not only sperm, but legal acknowledgment of the child. They referred to Mr. X in a kind and gentle way, expressing their gratitude for his service.

Over lunch Kaoru and Lil told me about the dramatic six months they had spent finding a suitable donor and then working with local authorities to get permission to create a *koseki* for their child upon birth. Everything seemed to be going smoothly for the couple in early 2014, including the pregnancy, when they went to inquire at the local city office about the registration of paternity. There they were informed that one of the conditions of Lil registering Mr. X as the father of her child was that she was required to declare that she was not married to anyone else anywhere. This condition of the notification process and the acknowledgment rule assumes that anyone who is pregnant must be heterosexually involved with a man. The proof of non-marriage was required to assure the authorities that one man's child would not be acknowledged by another man. Beneath these rules is the assumption that an unmarried pregnant woman must be a lover/extramarital partner of a married Japanese man. At its core, it assumes all pregnant women must be heterosexual, must have a male partner whom they cannot marry because he is already married to someone else; in other words, the structure of good wife (married)/bad woman (unmarried and sexually active) continues in the law and *koseki* registration procedure. Since Lil and Kaoru are for all intents and purposes married in Great Britain, these heteronormative assumptions built into the law did not apply. Lil told me that she did not think the British government would be willing to classify her as an unmarried woman. I questioned Lil on this assumption and she assured me that her civil union would show up on her identification information and that she would not be able to get a letter of proof of non-marriage from the British government. Her status as already married (even though it was within the context of the heteronormative rules that assume already married equals already married to a man who should have first rights as father) made it impossible to give a third party the right to be father and thus the notification of acknowledgment of paternity before birth (a type of acknowledgment known as *taiji ninchi* or *in utero* acknowledgment) was declined by the *koseki* official at the local city office.

Too married, but not married enough

Kaoru told me that when she and Lil were told that because they were already married, they could not have an additional person named as a father of their child, they were dismayed. Caught in a catch-22, unqualified to be considered married in Japan because they were a same-sex couple, and thus unable to grant citizenship rights through Kaoru, they were also considered too married to be able to have another adult named on the *koseki* in order to register his paternity. When they tried to file an Acknowledgment Notification form, it was declined because the mother, Lil, was already married. If they had not registered their marriage, a civil union, in Great Britain, Lil would have been able to have the donor register his paternity and guarantee citizenship and acknowledgment of paternity, though not legitimacy, in the *koseki*. But as a married person with a spouse who could not be recognized as a spouse by Japanese law because of the same-sex fact of their relationship, they were left in a precarious situation, too married to have the father's role acknowledged and too un-married to permit citizenship through the Japanese citizen, Kaoru. The civil union recognized under British law, which was not recognized in Japan as a marriage, made the particular registration of Mr. X's paternity impossible.

Despite the setback, Kaoru and Lil were determined to pursue alternatives to allow their child to acquire Japanese citizenship. They explained their situation at the city hall where the *koseki* administrator working on their request showed a certain amount of compassion and agreed to make a call to higher authorities to request permission to at least submit a notification of acknowledgment for consideration in their unusual case. However, after waiting in the office for several hours they were told to return home and await further notice. Kaoru told me that it was nearly three months before they received an answer from city officials. "It was a long process and we were worried the whole time." When they finally called from the local office and reached Lil by phone, they told her that the father's paternity could be registered so that the child could be granted Japanese citizenship. They were hugely relieved and described the sense of achievement they had from working around the system to earn rights for their child that current laws, rooted in the assumption that all marriages include a male who will be assumed to be the father, exclude. Although the child could not be registered in Kaoru's *koseki*, he/she would gain citizenship and *koseki* registration. In the case of their child, the *koseki* would have to be headed by the infant. There could be no other head of household.

I asked if they had received any formal letter confirming the acceptance of their request, trying to imagine what it might have said and how the decision would have been rationalized, but Kaoru, looking rather droll, said "no, we received nothing in writing." I asked what the basis for the decision was and they admitted that it was unclear. Kaoru mentioned the 2008 Supreme Court decision granting citizenship to children fathered by Japanese men and born out of wedlock to Filipina, non-citizen mothers. This Supreme Court decision overturned a lower court decision that had denied the plaintiffs' claims for Japanese nationality.[4]

Kaoru and Lil's child had to be registered in a *koseki* alone because Lil, as a non-Japanese, could not register herself as a parent of the child in a *koseki*. Although Lil considers herself married to Kaoru, her child is considered born out of wedlock (*chakushutsu de nai shi*) by Japanese law, because she is not married to a Japanese man. As discussed in Chapter 3, a child born to an unmarried couple must be registered in the mother's *koseki*. If the mother is non-Japanese then it is not expected that citizenship will be claimed. Lil could not marry the donor (even for a day through rapid registration of first marriage and then divorce) because she is already married to Kaoru in England. Therefore, the child needed to be acknowledged by the father and have an individual *koseki* with him/herself as *hittōsha*, first registrant. On such a *koseki* the mother's name is written in the margin (because she is not Japanese) and the father's name is also written in the margins (because the child was born out of wedlock). Kaoru, the other mother, is not named as a parent on the child's *koseki* because there is no way to acknowledge her as a legal or biological parent.

When I Skyped with the couple and their now three-year-old child while in Japan in July 2017 I learned that the child had been given Kaoru's last name, linking the child very clearly to her. Giving the child Kaoru's surname does not represent any legal relationship between Kaoru and the child. However, it is a symbolic and powerful way to work within the law and to acknowledge the (non-legal, but social and kin) relationship between mother and child. Only the biological parents are named on the child's *koseki* and neither has the same name as the child.

Kaoru said that she found the individual registration for the baby ironic given that all Japanese children have to be registered in somebody's *koseki* since *koseki* are based on the family, not the individual, and no baby comes from nowhere. A *koseki* with only one person is possible and indeed becoming more popular as people stay single longer and decide to separate from their parents' *koseki* using a process initiated with a notification called *bunseki*.[5] However, in her child's case, an individual *koseki* was the only way to register the child at all. An individual *koseki* for an infant is extremely rare, basically limited to cases where the parents are unmarried and the mother is a non-citizen who has requested and received permission to grant Japanese citizenship to the child.[6] Although a father or mother might be named on numerous different birth certificates in the US registration system, in Japan each citizen can only be listed on one *koseki*. That structure is fundamental to the way the *koseki* operates. Through Lil and Kaoru's experience we can understand certain constraints in the *koseki* system that appear when a family departs from the expected norms of family structure backed by the Civil Code and the *koseki* system.

Uncertain of where their careers might take them in the future and where they might live with their child, Lil and Kaoru were determined to help their child get citizenship through *koseki* registration. As Lil said in our interview, "we are living in Japan now and without a *koseki*, you can't get services such as health insurance, day care and public education." They were clearly concerned about the well-being of their child who would be raised in a bi-cultural household and while living in Japan, at least, would need to have some proof of residence and legal status. "I only have residency status through my job. When I stop working to take care of the baby, it is really important that the baby has some basis for residency other than me." The catch here is that although Kaoru and Lil intend to raise their child together as a family, Kaoru and Lil have no recognized proof of relationship, making Lil appear as a foreign single mother without a claim to live in Japan or receive benefits from the state for her child. Kaoru and Lil are trying to register their family which is out of the bounds of the *koseki* system. The fact that they were able to obtain citizenship status (and *koseki* registration) for their child may suggest flexibility. However, ultimately the system was highly rigid and without the expertise and support from other activists and a good deal of perseverance it would have been impossible for them to obtain citizenship status for their child. Other couples in a similar situation might be more likely to leave Japan or accept the foreigner status of their child. Kaoru and Lil's manipulation of the system to gain citizenship for the infant shows the kind of last option scenario that many of the activists I've interviewed faced. If they had not succeeded in convincing the sperm donor to acknowledge the child as his own, then surely the child would have been outside the *koseki* and would have lacked citizenship. From the couple's point of view a child who is half-Japanese with Japanese citizenship is much less likely to face discrimination and ostracization in Japan than one lacking citizenship status. They hoped their strategy would pay off in their child's experiences living in Japan as a citizen, rather than a foreigner.

Intentional and unintentional mis-registrations

By acquiring Japanese citizenship for their child, Kaoru and Lil manipulated a system that through its structure attempts to keep a child like theirs off of the *koseki* and off the rolls of Japanese citizenship. Their creative strategy, which included sperm donation from a man willing to also acknowledge his paternity (rather uncommon in sperm donation), helped squeeze their child into a column of a *koseki*, therefore reaping the vast benefits of Japanese citizenship. Other people I interviewed have used creative mis-registrations or misrepresentations of their relationships to either gain benefits or resist discriminatory designations on the *koseki*. Women trying to keep their surnames employed numerous strategies to subvert and work around the hegemonic single-surname rule.

Creative legal workarounds to reduce the effects of discriminatory laws and notations on the *koseki* mark the lives of many of the activists I interviewed. In the face of the common surname law (Civil Code 750) they have settled for common-law marriage or have married and divorced in order to resume their maiden names. In some cases, they have stayed legally married, ignoring expectations that they use their husband's names. The few women I interviewed who married foreign men were able to keep their maiden names by heading their own *koseki* (since their foreign husbands could not be listed on the document and thus the choice of which name to assume in marriage was quite different than when the husband was Japanese).[7] However, it is worth noting that it is possible for a Japanese woman married to a foreign man to change her name to her husband's name. A name-change notification is used so that she, as first registrant on the *koseki*, has a foreign name written in *katakana* to represent the Japanese pronunciation and "non-Japanese" provenance of the name. For the most part, wives in international marriages registered their children under their natal surnames on the *koseki* because of the convenience.[8] The Civil Code and Koseki Law create numerous barriers for women who want to keep their maiden names in marriage, as discussed previously. But underlying the conservative family laws are gender understandings and social practices that have been extremely slow to change, and most of the people I interviewed well understand this. Given the glacial pace of change to the Civil Code and Koseki Law, activists have created ways to avoid discrimination and gain benefits through various mis-registrations and workarounds to improve their lives in the here and now. Creative mis-registrations or non-registrations provide a potential space of resistance to the rules of the *koseki* system and an examination of these workarounds can help us understand the feminist work to dismantle and expose the contradictions in the system.

Misrepresenting actual relationships

The problem of misrepresentation of actual relationships has been an ongoing theme in my conversations and interviews with feminists who strive to represent their relationships accurately and honestly on the *koseki*. Certain kin relationships like same-sex marriage or common-law marriage simply cannot be represented on the *koseki* at all. My interviews and conversations often returned to unusual strategies employed in order to use a maiden name or re-take a maiden name. Most of the people I interviewed in common-law relationships had chosen to stay legally *un*married in order to keep their maiden names. But for many of them, it was an inconvenient choice. More than once I was told, "*kekkon shitai to omotta, kedo namae ga*" ("I wanted to get married, but for my name"). As discussed in earlier chapters, common-law marriage makes it possible to maintain separate surnames, but couples who are not legally married cannot use the *koseki* when it would be helpful, precisely when gaining benefits derived from family membership or marriage.

For many interviewees I spoke with during the year who live with their long-term male partners and consider themselves married, the relationship between husband and wife is invisible on the *koseki* and this invisibility is a cause for some concern.[9] For some common-law couples, like those in the *fūfubessei* lawsuit discussed in Chapter 2, common-law marriage creates social and financial problems that spill over into household plans and fulfilling obligations and expectations of employers or families. In the Valentine's Day lawsuit over the right to separate surnames, two plaintiffs were told they were ineligible for a home loan because they were not legally married. For a third plaintiff in the case, Tsukamoto-san, a school teacher, there was pressure to be legally married and use her husband's surname because of her role as a public servant and role model for children. As a woman born before the end of World War II she does not consider herself a marriage resister but rather a surname defender. Couples in common-law marriages who have been together for decades and have children together consider themselves married, but have no standing in the courts should something unexpected happen that might include the division of property or parental rights to child. Healthcare, social security, and other social services made available through the government programs recognize individuals in common-law marriages as single. To date, there is no legal status of common-law marriage despite the popularity of *jijitsukon* among a growing number of couples. Equally importantly, employers who are often the conduit for social and economic benefits, including housing and insurance subsidies, relocation stipends and family stipends, generally do not recognize a common-law marriage since a copy of a *koseki* is often the basis for proving marital status and family size, making one eligible for benefits. At work, marriage may be an asset or a demerit for women these days, depending on the office climate and other social expectations, so it is difficult to know if the appearance of being single is a help or hindrance in a woman's career, but common-law marriage makes it impossible to show that a relationship is real and long-term.[10]

In addition to common-law marriage and the apparently single adults it creates, the practice of paper divorces, perhaps convenient, misleads anyone looking at the *koseki*. Given that divorce is considered a blemish on a *koseki*, a woman who repeatedly divorces and marries the same man (all the while his name stays the same) may be achieving her desired goal of acquiring legal use of her maiden name, but she is hardly protecting the good name of her family. Instead, something looks quite amiss on the material *koseki* document that references numerous past divorces. Like the other workarounds to keep a maiden name, this workaround does not come without a price. Furthermore, in several cases, we can see that a paper divorce that had been arranged in order to re-take a maiden name, later resulted in an actual separation and divorce. Two interviewees told me that it was easy to actually divorce several years after their paper divorces since there was no legal marriage in place anymore. In both cases, the women I interviewed made it clear that their husbands never fully understood why their wives had chosen to re-take their names and officially divorce by submitting a Divorce Notification at the municipal offices. This suggests that paper divorce, which works for some couples, depends on good communication and a husband's understanding of his wife's desire to keep her name.

Non-marriage or divorce and re-marriage are strategies that result from a desire to keep separate surnames in marriage, or more precisely for a woman to keep her maiden name. The effects of these strategies and workarounds, however, lead to unintended consequences that become new *koseki* problems. The designation of "illegitimacy" for children of common-law marriages has vexed surname activists for several decades and to date discriminatory notations have not been completely eradicated. Because they have not been legally married, the women who have children with common-law husbands appear to have children born out of wedlock, as discussed in Chapters 3 and 4. The stigma to both mother and child continues to shape critiques of the *koseki*. The changes to the *jūminhyō* and *koseki* in 1995 and 2004 respectively significantly reduced discriminatory details on the *koseki*, and yet the problem of appearing as a single mother in order to keep a surname creates multiple problems in daily life because children in such situations are assumed to lack fathers in a society where fatherlessness is still a social handicap.

Non-registration of a child in order to avoid the mark of illegitimacy may look like a case of cutting off the nose to spite the face. Certainly, when I heard Tomizawa-san's rationale for leaving her son off the *koseki* for so many years I was astounded. But as she well knows, this kind of resistance to the system actually helped bring about the changes that took place in the *koseki* designation of children born out of wedlock in 2005. Non-conformity to *koseki* norms is hardly an easy choice in Japan today. But activists who stand their ground and demand non-discriminatory notations on the *koseki* and notification forms are slowly making progress. The price of non-registration of a child includes a sense of risk and insecurity in daily life. But once saddled with the mark of illegitimacy in a *koseki*, one can never fully escape it, even if it is known that the parents were merely keeping separate surnames. Compared with these strategies to avoid certain types of blemishes and discrimination on the *koseki*, I think we can understand Kaoru and Lil's actions as an attempt to gain access to certain privileges of citizenship that were not likely to be granted to a family in a relationship not recognized under Japanese law.

In addition to the creative workarounds to either avoid discrimination or gain benefits of citizenship and residence discussed here, there are other types of misrepresentations of family structures that also expose bias and contradiction fundamental to the koseki system.

During the past several decades international marriage *kokusai kekkon* 国際結婚 has been on the rise. I interviewed three people in international heterosexual marriages and discussed the issue of international marriage with many others. *Kokusai kekkon* 国際結婚 (international marriage) appears on the koseki as very close to non-marriage. For a married couple, as in Figure 5.1, the head of household is listed with full birth and parental information followed by the spouse of the head of household. This format is universal. It is understood as the way to represent a married couple. However, as should be clear, in the case of same-sex couples or international couples there is no spouse named in the spouse location. And thus, because the spousal column is ubiquitous, when a spouse is not listed in the expected location it appears that the head of household is a single adult.

The *koseki* in Figure 5.2 shows a *koseki* headed by Otsuno Umeko. There is no one named in the spousal column, so a quick look suggests that Otsuno is single. However, in the details written above Otsuno Umeko's name, the fact that she is married to Wayne Van den Bosch is noted. Given that all koseki for

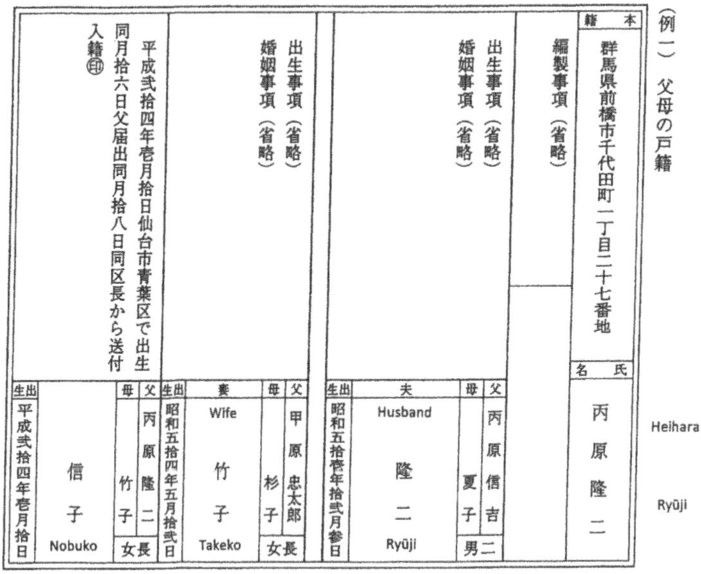
(例一) 父母の戸籍
本籍 群馬県前橋市千代田町一丁目二十七番地
氏名 丙原隆二
Heihara
Ryūji
編製事項（省略）
出生事項（省略）
婚姻事項（省略）
父 丙原信吉
母 夏子
二男
夫 Husband 隆二 Ryūji
生出 昭和五拾壱年拾弐月参日
出生事項（省略）
婚姻事項（省略）
父 甲原忠太郎
母 杉子
長女
妻 Wife 竹子 Takeko
生出 昭和五拾四年五月拾弐日
平成弐拾四年壱月拾日仙台市青葉区で出生
同月拾六日父届出同月拾八日同区長から送付
入籍㊞
父 丙原隆二
母 竹子
長女
信子 Nobuko
生出 平成弐拾四年壱月拾日

Figure 5.1 Marital *koseki*, international marriage.

Japanese citizens list the spouses side by side, this form of notation has a distinct visual and symbolic effect. While the *koseki* activists I interviewed were not terribly concerned about this, in part because the women were able to keep their own names when married to foreigners, it is a topic that receives much attention in the community of international marriages, particularly on websites for foreign spouses of Japanese who feel marginalized in Japan as non-citizen parents of Japanese children.

The absence of information in the spouse column, the expected location for spousal name and information, particularly given the graphic nature of the document, makes such a *koseki* appear askew. The columns for spouse are there, but no name is in that location, suggesting that there is no spouse. Because of its material form and its symbolic power as a representation of family, this apparently non-marital *koseki* reveals the expected norms and the divergence from those norms that an international marriage represents.[11] For some people in international marriages, the marginal notation of a foreign spouse on a *koseki* is not necessarily perceived as a major problem. However, when a foreign spouse, husband or wife, becomes the only surviving spouse, or becomes the primary parent when complicated kin matters develop, the individual's marginal status on the *koseki* can be amplified and result in serious consequences related to parental rights, inheritance rights, and so forth.

For Japanese women who marry foreign men, nationality works in their favor on the *koseki*. Rather than being expected to give up their names for their

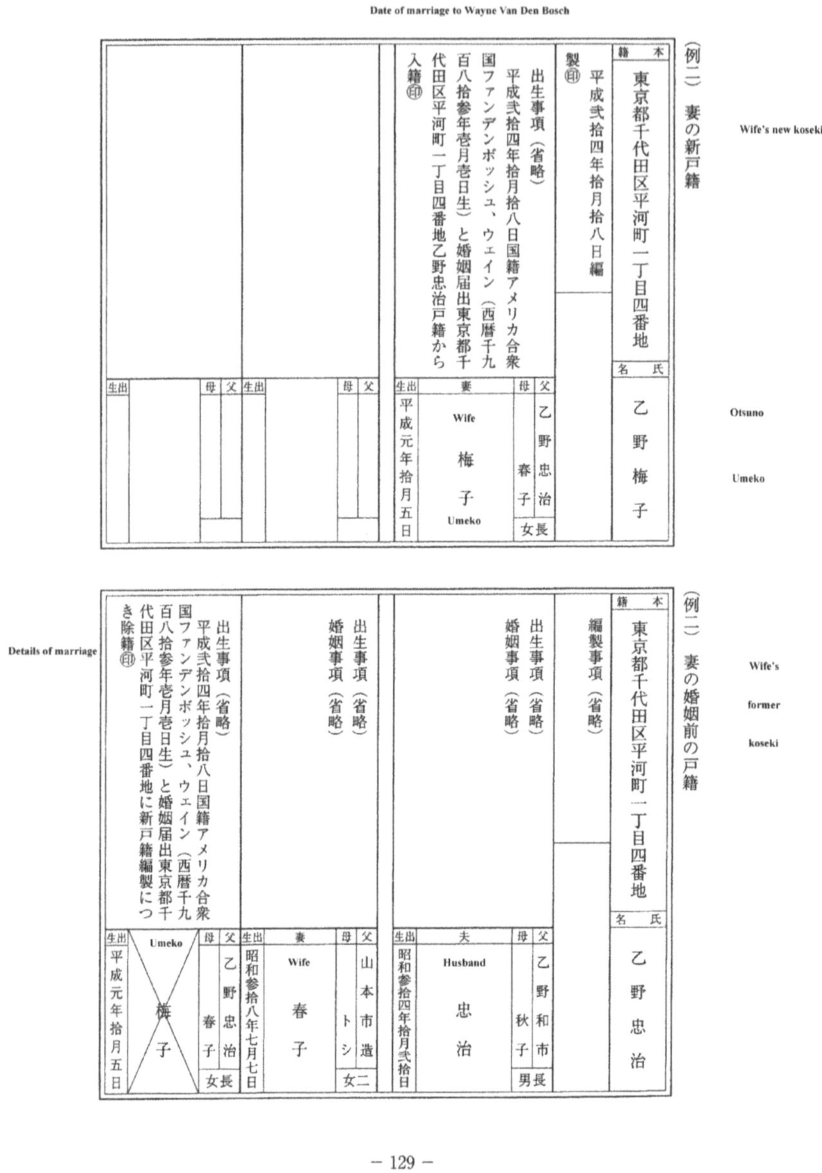
Date of marriage to Wayne Van Den Bosch

（例二）妻の新戸籍

Wife's new koseki

本籍 東京都千代田区平河町一丁目四番地

氏名 乙野梅子

Otsuno Umeko

平成弐拾四年拾月拾八日編製㊞

出生事項（省略）

平成弐拾四年拾月拾八日国籍アメリカ合衆国ファンデンボッシュ、ウェイン（西暦千九百八拾参年壱月壱日生）と婚姻届出東京都千代田区平河町一丁目四番地乙野忠治戸籍から入籍㊞

父 乙野忠治 母 春子 長女

妻 Wife 梅子 Umeko

生出 平成元年拾月五日

（例二）妻の婚姻前の戸籍

Wife's former koseki

本籍 東京都千代田区平河町一丁目四番地

氏名 乙野忠治

編製事項（省略）

出生事項（省略）

婚姻事項（省略）

父 乙野和市 母 秋子 男長

夫 Husband 忠治

生出 昭和参拾四年拾月弐拾日

出生事項（省略）

婚姻事項（省略）

父 山本市造 母 トシ 女二

妻 Wife 春子

生出 昭和参拾八年七月七日

Details of marriage

出生事項（省略）

平成弐拾四年拾月拾八日国籍アメリカ合衆国ファンデンボッシュ、ウェイン（西暦千九百八拾参年壱月壱日生）と婚姻届出東京都千代田区平河町一丁目四番地に新戸籍編製につき除籍㊞

父 乙野忠治 母 春子 女長

Umeko 梅子

生出 平成元年拾月五日

– 129 –

Figure 5.2 Parental *koseki*.

husbands, they are expected to maintain their Japanese maiden names. As *hittōsha* of their *koseki*, they can easily make changes to their household record without interference from a husband or husband's family. As Japanese citizens married to non-citizens they are assumed to be the head of the household and their family name is assumed to become the name of the offspring who will be entered into the "mother's" koseki upon birth. Most difficulties related to the registration of an international marriage arise if the marriage dissolves. Divorce between two Japanese citizens can follow an expected path toward a decision for one parent or the other to maintain sole custody of children, though it is most common for the husband to keep the children in his *koseki* even if the wife has custody I've been told. Leaving the children in the father's *koseki* makes it easy for them to maintain their surname and their legitimacy status and it allows a divorcing mother to perhaps re-marry or return to her parents' *koseki* or even begin a new koseki of her own through the process of *bunseki* (separation of *koseki*). But custody and name matters are quite different for Japanese and non-Japanese.

In her discussion of international marriage, the historian, Kamoto Itsuko, notes that foreigners were considered Japanese upon marital registration in a *koseki* during the Meiji period (Kamoto 2014). Edict 103 of the Grand Council of State (*dajōkan*), known as Regulations on Japanese-Foreigner Marriage, issued one year after the Koseki Law went into effect, created a system of immediate naturalization for marital partners of Japanese nationals. As Kamoto points out, the regulations in Edict 103 were issued before the Meiji government had created a Civil Code, a Constitution, or a Nationality Law. This is surely evidence that the practices of the *koseki*, often delineated through government edicts like 103 provided the framework for subsequent family law that later became incorporated through the Civil Code, first in 1889 and then in 1898 (Chapman and Krogness 2014). In 1873 when the Edict on international marriages was issued, the legal definition of nationality had not been precisely defined, due in part to controlled contact between foreigners and Japanese at Dejima during much of the Tokugawa period (1600–1868). When the rate of international marriage started to increase in the mid-nineteenth century, Japanese nationality was granted to a foreign spouse upon marriage. To be married to a Japanese meant that (within Japan, at least) one became Japanese, even if born French, American, or something else. The Nationality Law of 1899 clarified a policy of paternal *jus sanguinis*. Until the law of *jus sanguinis* went into effect, it was easy to see that inclusion as a family member recorded on a *koseki* made one a citizen and not the other way around. However, the granting of Japanese citizenship upon marriage did not extend into the twentieth century.

The representation of international marriages on *koseki* today shows that marriage no longer confers citizenship. Mapping of information onto particular locations on the *koseki* provides a quick and dependable way of reading family or kin relations. Almost like checking someone's ring finger for evidence of marital status, a quick look at a *koseki* shows that a spouse is listed or not and so in the case of the international marriage, unless the spouse has naturalized as Japanese, there is a void where a spouse should be named. While many Japanese in international marriages accept this situation without complaint, others argue that it puts the foreign spouse in an inferior relationship to the Japanese spouse and complicates power dynamics within a couple. The number of international marriages grew steadily in the late twentieth century, topping out at 40,000 in 2006, a peak from which it has receded in the past decade. Currently the rate of international marriages is 3.25 percent, down from 6.25 percent in 2006 (Ministry of Health and Welfare).

Like all marriages in Japan, international marriages are specified through Koseki Law and the Civil Code. When I interviewed the head of the *koseki* division at the city offices in Musashino City, a district within the greater Tokyo metropolitan area, I was told that the reason that foreigner spouses are not listed in the column for husband or wife on a *koseki* is that the *koseki* is designed to track Japanese people. "*koseki wa, nihonjin no tame desu yo*" ("The *koseki* is for Japanese people").

The irony for the *koseki* today is that people who are in legal kin relationships must, by law, be on the same *koseki* and yet foreigners are not noted as family members on *koseki* even when they are married to Japanese nationals. As a tool for household registration, then, the *koseki* is expected to record family members regardless of nationality, but as a record of citizens, foreigners do not belong. That is why, I suppose, I was told that there is no problem with the foreign spouses being noted in the margin of the *koseki* "because the *koseki* is for Japanese." Clearly, the *koseki* system is stretched as it tries to incorporate and handle the many idiosyncratic situations brought about in a changing globalized society.

Unlike couples in international marriages, same-sex partners cannot even have their names listed in the margin of the *koseki*. Like same-sex marriages, international marriages cannot be contained or represented accurately within the lines or sections of the *koseki* as it exists today. Obtaining proof of relationship in order to claim that the child of a marital partner is one's own (taken for granted in heterosexual marriages and in Japanese law, Article 772 of the Civil Code) is nearly impossible in Kaoru and Lil's case. During my conversation with Kaoru, she said, "I wonder if I will have problems picking up our child at daycare or school, since the child will not have a legal relationship with me?"

In order to acquire citizenship for their child, born within a marriage unrecognized by Japanese law, the couple used various laws to find a workaround. If same-sex marriage were allowed in Japan, a baby born to a foreign spouse would have citizenship rights through the Japanese spouse. In the absence of a legally recognized marriage in Japan, the presumption of spouse's role in parentage is nonexistent.

Misrepresenting same-sex relationships and workarounds

Because same-sex marriages cannot be documented on the *koseki* and adult adoption is an easily achieved status, some same-sex couples have employed adoption to obtain family membership (Maree 2004; Mackie 2014). While not a widespread or typical process, same-sex partner adoption provides a way to legally claim a kin relationship when the more precise status of marriage is unavailable. Adult adoption has a long history in Japan, used widely for bringing in a head of household to a family lacking a qualified heir. Adult adoption (generally employed to help the adopter's household, rather than to help an adoptee) is more widespread than infant or young child adoption. Hertog (2009) discusses the high frequency of adult adoption, rather than child adoption, as linked to "the moral imperative to raise one's own child" in Japan. Her research showed a cultural preference for abortion over adoption, where adoption is perceived as "a solution morally inferior to abortion" (Hertog 2009: 34).

Adoption of same-sex partners is a legal strategy that is not unique to Japan. But the lack of stigma surrounding adult adoption may make it slightly easier to use. Article 793 of the Civil Code specifies conditions for adult adoption, which is limited only by the requirement that the adopter be at least one day older than the person adopted.[12] I asked Kaoru and Lil if they had considered adoption, knowing that other same-sex couples had used this as a way to gain a variety of rights and benefits within the Japanese system. Kaoru immediately said that she thought it was a weird idea. Adoption makes no sense for us, we are already essentially married. We have a civil union. They both said they do not want, even if only on paper, the appearance of a mother–child relationship when they are partners. In addition, Kaoru mentioned her hope that someday same-sex marriage would be legalized in Japan, she went on, "how would one go about marrying their own mother?"

According to Maree (2004) common problems for unmarried couples can be solved through adoption, such as proof of kin relationship in case of emergency or medical crisis or inheritance (Maree 2004). But adoption to gain the legal rights of kinship, like the other strategies I discuss in this chapter, is really born of the rigid and narrow definitions of marriage and family that continue to shape the choices and options within the tight, heteronormative family structure centered on the Meiji-era Civil Code and Koseki Law. When common-law couples appear to have children born out of wedlock, when married couples appear divorced so that the wife can reclaim her maiden name, and when married same-sex couples in one country appear to have no relationship in Japan, the *koseki* system's limitations become visible. Many of the individuals I have interviewed for this book have spent years and years of volunteer labor to bring these often invisible rules and limitations of the *koseki* to light in order to raise awareness of the imposition of the state into individual life choices. Same-sex marriages, like long-term common-law relationships, simply cannot be documented adequately on the *koseki*. It is precisely the inflexibility and standardization of the family model on the *koseki* that make variations to the norm so noticeable. A blank in the spousal column, so obvious and visible on the physical *koseki* document, suggests that a person is single when they might be in a common-law marriage, an international marriage, or a same-sex marriage.

Some of the inaccuracies and misrepresentations on *koseki* discussed in this chapter are intentional, forms of subverting an oppressive social structure in order to avoid women's coerced name changes or other effects of a heteronormative and patrilineal system. Other misrepresentations are unintentional and unavoidable. Structural bias in the *koseki* system, the Civil Code, Koseki Law, and the document itself, favors certain people, including men who generally are named first on the *koseki* as a matter of deeply held attitudes at the root of Japanese male privilege, citizens who marry other citizens, and obviously, people in heterosexual relationships rather than people who are in same-sex relationships. Despite the elaborate governmental bureaucratic systems designed to keep track of everyone and make sure that all notifications in registrations are accurate, resistance to antiquated gender norms and the development of new family forms have created messy and inaccurate *koseki*.

David Chapman argues that the 1871 Koseki Law, despite being implemented in order to create social order from social chaos in the process of Japanese modernization, in fact, produced a new kind of chaos, of "strangers" and "undecidables," people who did not neatly fit within the families of the nation specified in the Koseki Law (2014). In a related vein, I argue that *koseki* are further disordered and chaotic as a result of efforts to create new kinds of families in recent generations that do not adhere to the strictures of patrilineal heteronormative families. The resistance of activists and others whose families don't conform, create a new level of inaccurate and chaotic *koseki* that cannot be relied upon for true representation of family. The separate-surname activists, through their non-registration of marriage, non-registration of children born out of wedlock, and repeated registration of marriage and divorce, misrepresent the true status of their relationships in order to avoid discrimination. These resistances disrupt *koseki* logic and undermine *koseki* claims as the official "accurate" census of the Japanese people. Inaccurate *koseki* created by activists resisting the same-surname rule do not reflect the true status of individuals and families and have the effect of distorting statistics based on them.

An old friend with whom I informally discussed my research in the first weeks of my research in 2012, told me that it does not really matter if a *koseki* does not accurately reflect a family. "We know who is in our family." "As long as we feel like family, if the *koseki* is different, it doesn't matter." As she said, people know who makes up their family regardless of what the *koseki* shows. However, the *koseki*'s meanings and relevance come at multiple levels. For individuals and families, they can be a source of discrimination or something to take for granted and ignore. But as a kind of census of citizens, marriages, and kin groups, inaccuracies created by activists surely cause problems. Not everyone agrees on the importance of the accurate portrayal of the family on a *koseki*.[13]

For some people, the benefits and legal provisions afforded through family membership or citizenship matter deeply. People like Kaoru who cannot grant residence or citizenship to her child or her partner feel terribly marginalized by the laws and system that exclude a same-sex marital partner. Because the government grants rights and duties based on *koseki* membership, the representation of a marriage or a family on a *koseki* can have significant monetary, administrative, and social effects. So, although not everyone cares if their *koseki* is accurate, citizens must function within bureaucratic systems predicated upon the assumption that what is on the *koseki* reflects true kin relationships.

Conclusion

Koseki activists have worked to increase awareness of the biases in the family system by allowing absurd notations to be made about themselves and their own children. They have challenged discriminatory terms in the *koseki* system, refused to have their names changed in marriage, which has led to their actual family structures being misrepresented, appearing single, when in fact married, or appearing divorced when actually married. They have critiqued bias based on gender norms and kept their maiden names only to find their children considered illegitimate and their marriages unverifiable. They have sued their local and national government, including the national lawmakers for failing to eradicate discriminatory language in the Koseki Law and Civil Code. It is through these myriad techniques that they have undermined and subverted patrilineal and patriarchal *koseki* logic even as they have been inconvenienced by their own actions.

As I have shown in the preceding chapters, nineteenth-century notions of gender, family, and human sexuality that became building blocks of the Koseki Law of 1871 continue to shape understandings of family and Japaneseness today. For many people, a shared surname defines family and provides a link to the Meiji-era *ie seido* which has been widely understood as essentially and fundamentally Japanese and "traditional," despite evidence that it was being produced in a new form derived from new Meiji-period laws, and in particular, the Koseki Law, in the rapidly modernizing period of the late nineteenth century. Activists trying to maintain their surnames and become legally married struggle with the constraints of Meiji values and expectations as they navigate their lives in the twenty-first century. In this sense, they simultaneously contend with aspects of out-of-date gender norms laid deep in the law and *koseki* document and the post-industrial, late capitalist society where the individual and gender equity are purportedly highly valued and even guaranteed in the Constitution. Today, young women are expected to aspire to high achievement in education and career befitting the late-twentieth and early twenty-first-century gender norms, a far cry from what was expected or tolerated back in Meiji or even in the early postwar period in the mid-twentieth century. As a result of the contradictions between family law and lived experience, many of the activists I interviewed for this book have resisted the Koseki Law and developed workarounds to avoid discrimination and to name themselves and their children in ways they deem appropriate. These misrepresentations expose limitations of the *koseki* system and document and force us to ask why the government is determined to keep the *koseki* system in place more than 70 years after the end of World War II. In an ever more interconnected world with more kinds of relationships and marriages and a blurring of national borders, what is the purpose of keeping family registers designed for only Japanese citizens when increasingly families in Japan do not consist of only Japanese citizens? What is the point of forcing married couples to register their small families under a single surname when the legacy of patrilineal name preference and patriarchal family structure weighs so heavily on the possibility of choice?

Misrepresentations produced through resistance to the *koseki* system expose deep contradictions in a structure that is increasingly out of step with the fluid nature of Japanese kin groups, where divorce, re-marriage, out-of-wedlock childbirth and non-marriage are all common occurrences. These disparities between lived kin relationships and representations of relationships on the *koseki* expose gaps between law and practice, gaps between expectations of normative family life and evolving family forms that are on the rise in early twenty-first-century Japan. The *koseki* system itself straddles the Meiji past and the twenty-first-century society. Its form and function were borne of Meiji understandings of not only the *ie seido*, but also gender norms and expectations and the near-compulsory system of heterosexual wedlock at the time. In this sense, the *koseki* document can be seen as a remnant of bygone days, temporally stretched to the twenty-first century where new social contexts and changing norms of gender and sexuality explode. The Meiji-era logic of the *koseki* structure bleeds through into the renovated but only minimally revised *koseki* document used in Japan today, revealing its desperate grasp for order that leads to chaos.

Notes

1 See Kamoto (2014) for an excellent discussion of the historical development of the modern *koseki* in Meiji at the same time that nationality law was taking shape as it was increasingly needed during rapid international contact after 1853.

2 As mentioned previously, Krogness has argued, the term, *jus koseki* (rule of the *koseki*) more accurately describes how Japanese citizenship rules have operated historically where inclusion on a *koseki* through marriage or adoption made one, de facto, a Japanese citizen (Krogness 2014). Lafcadio Hearn, Ogasawara Islanders, and others who are clearly not of Japanese descent (*jus sanguinus*) have been registered in *koseki* and deemed citizens (Chapman 2014). And under Japanese rule Koreans were registered in *koseki* of various types depending on where they resided and their colonial status (Kashiwazaki 2000).

3 See Mackie (2014) for a discussion of the sperm donation and IVF and *koseki* registration.

4 This particular decision came up during numerous conversations in my research because of the dilemma of non-citizen mothers who knew that the father of their child was a Japanese citizen and wanted to gain citizenship status for their children growing up in Japanese society. The 2008 Supreme Court decided in the favor of plaintiffs (children and their mothers) to grant their out-of-wedlock children Japanese citizenship. See Nobue Suzuki's excellent discussion of this issue and the Supreme Court decision of June 4, 2008.

5 *Bunseki* or separation of *koseki* makes sense in a variety of situations. By separating from a natal *koseki* a person has a bit more freedom from their parents and immediate family about when to marry, if to divorce, if to adopt, or in the case of being transgender, when to change gender legally. When in a parental *koseki* any of those life changes are visible on the parents' *koseki* and implicate family members in numerous ways, having the effect of suppressing certain non-normative life choices.

6 Usually one of the parents is a citizen and can head the household for the child. Otherwise, citizenship is not sought.

7 It is possible for a Japanese wife to take a foreign husband's name and use his surname on her *koseki*, but he will not be named as the head of the *koseki*, she will. Thus, it is possible to have a Japanese woman whose name is Jones as the head of

koseki, but it appears that most Japanese women married to foreigners register their *koseki* under their natal name even if they change their surname to their foreign husband's.

8 There are a number of different forms of naming possible in the case of international marriages. It depends where a couple lives and what they are trying to do. Children may have a Japanese surname on the *koseki* and go by their father's name in a foreign country where they live with their Japanese mother and foreign father. Similarly, Japanese women married to foreign men may decide to use husband's name abroad and their Japanese surname in Japan for myriad practical and symbolic reasons.

9 Clearly some of the people I interviewed were happy to be on the outside of legal marriage and on the outside of a marital *koseki*, like Tanaka Sumiko who rejected the whole system of marriage. Tanaka-san stood out in saying that she was not interested in a marital *koseki* even if she could maintain her maiden name because the *koseki* system reproduces the sexist and patriarchal *ie seido* system.

10 In Japan, marriage is widely perceived as an asset in employment for men. For women, a more complex set of gender expectations and social norms often complicate a career trajectory. Just a generation ago, tapping a working woman on the shoulder to recommend her retirement by age 25 was widespread in Japanese companies (see Kawashima 1995; Edwards 1989; Nemoto 2016).

11 *Koseki* structure is like a family tree. A quick glance at the visual representation of family shows the expected husband and wife and their children. One of the differences between a family tree and a *koseki* is the amount of detail that is discreetly associated with each named household member.

12 Adoption has been a common strategy for continuing a family line in the absence of an eligible heir, as discussed previously.

13 At the same time, inaccurate representations of family relations may not seem significant to people who perfectly fit societal norms. If they are heterosexual, married to another Japanese citizen, and neither cares about changing his/her name, then the *koseki* surely might feel benign.

Epilogue

December 20, 2017

On my most recent trip to Japan in July and August 2017 I met with many of the people who helped me with the research for this book, tying up loose ends and hearing updates about various lawsuits and activities. I spent a long afternoon with Tanaka Sumiko going over some of the things I had written about her activism and hearing about recent work she and Fukukita-san are carrying out with local municipal governments to encourage them to play a role in pressuring the national government to change current registration practices that discriminate against single mothers and children born out of wedlock. I also spent time with old friends at a recently opened drop-in center for women seeking a space safe from abusive partners and husbands; half day shelter, half community café and networking site, it was full of kids and women, sharing food and hanging out. The steadfast commitment and compassion that fuels the work of many of my informants and friends endures, as does the sexist system within which they toil. I gave several informants and friends the book of photographs cataloging the Women's March on Washington, January 21, 2017, *Why We March*, and we lamented the state of gender affairs, not only in Japan, but in the United States and elsewhere. But despite setbacks and challenges it was easy to see that Japanese feminists and feminisms continue apace.

As an anthropologist, I have been fascinated by feminist work to draw attention to the *koseki* system and to change it. The *koseki* system itself – its historical and ideological significance, its powerful gendered resonances across many spheres of society, beyond marriage and divorce, to employment, regional and national identity, nation and citizenship and, of course, heteronormative family structure – has been endlessly fascinating to study and there is much more to be done. But as I chatted with old friends last summer I continued to come back to the reality, one that fascinates them and me equally, that in the United States, where it is legally simple for a woman to keep her maiden name in marriage, a majority of women do not do so. This bald-faced fact sits at the center of my consciousness about my research on the problem of women's marital surnames in Japan. Does the US case suggest that even when the law is changed in Japan and women have a choice about their marital name they will hew to convention and take their husband's name? Or is something else at work? Does the high rate of name change in the United States suggest that gender ideology is stronger than legal rights to an equal status? What forces in American society keep women following a procedure – taking their husband's name – that often works against them, when they are required to change each form of identification, including bank cards and credit cards, and bear the burden of changing their names again should they divorce, which is not an uncommon result of marriage in this society? Furthermore, in the United States, even when women make the choice to keep their own names in marriage, they very rarely argue to give their children their surnames, settling for the gendered norm of giving the children their father's name. In other words, although we have the rights to do things differently in the United States, in more than half of all cases, women are changing their names to their husband's in marriage. Gender ideology and norms that have shaped how people understand themselves in society play a significant role here and we can assume they will in Japan even when the law is changed.

This is not to argue against a change in Article 750 of the Civil Code, rather to warn that a legal change alone will not provide the panacea that some hope it will. In an ironic case, a prominent businessman, Aono Yoshihisa, president of a software company, who happened in an unusual twist to take his wife's name in marriage, plans to sue the government over the surname issue despite the 2015 Supreme Court decision that Article 750 is not unconstitutional. Could a male plaintiff, the reverse of the usual plaintiff and critic of name-changing rules, be the one to finally get a fair hearing in Japan? It would not be a surprise. From the start, the problem of the *koseki* and the single-surname rule has been about perspective. From a male point of view, a wife having to change her name upon marriage is not a problem. Wives are understood as entering husband's families and the name change is just part of that larger understanding of marriage. But when a husband has changed his name and the change begins to chafe and he is a successful businessman, well, it just might be time to change the law.

In the United States right now we are seeing an historic examination of structures of racism, sexism, and masculine privilege, systems that have warped and shaped American institutions for centuries. Women who have been silent and silenced for decades on the matter of sexual harassment and rape in the workplace are speaking out about their experiences in Hollywood and in Congress, in factories and in boardrooms. The Black Lives Matter movement, which started in 2013 in response to the acquittal of a man who shot an unarmed African-American high school student and came into full force in response to the police shooting of Michael Brown in Ferguson, Missouri in 2014, has revealed the patterns of racism that have shaped American life for centuries, exposing the ways that the shooting of innocent black children has been normalized through systemic portrayals of African-American males as violent and dangerous when, indeed, it is white institutions, individuals, and values, like ideas of racial superiority, that have been violent and dangerous for centuries.

In Japan, sexism and racism operate in different registers through different ideologies and practices, as the *koseki* system reveals. When trying to make sense of the challenges Japanese women face, I have tried to show that they are situated differently in a structure that is a different kind of patriarchy than the type known in the United States. There is much less discussion of liberal values, individual freedom, liberty, or even justice in Japan, where ethics are more often understood through social context and situational relationships. To be ethical, for example, may mean to not defame one's family name, one's employer, or the nation. It is a worldview and a mindset that shapes the behavior of even some of the most ardent feminists determined to crack the hierarchical and sexist systems of their society. But despite these differences in some of the framings of core values, women and men who would like to move a bit more freely in society with a bit less stigma attached to their life choices are determined to keep pushing for social justice and equality for themselves and others.

It is difficult to end this monograph on a note of hope. So many unwavering and determined activists have struggled for so long and yet they continue to be silenced and marginalized within the public sphere in Japan. Still, to quote the American anthropologist Margaret Mead, "don't doubt that a small group of thoughtful, committed people can change the world. Indeed, it is the only thing that ever has." In humble awe and respect for the women and men working to eradicate gender bias in the *koseki*, I offer this study of the formidable work of a small group of thoughtful and committed people who are changing Japan.

Appendix I

Association (*Kōryūkai*) timeline

1988	May 9: Tanaka and Fukukita file lawsuit against Musashino City and the Mayor of Musashino City for discrimination against their child. October 22: Association in Support of the Lawsuit Against Discriminatory Expressions in the Resident Register formed.
1992–1993	The Association in Support of the Lawsuit Against Discriminatory Expressions in the Resident Register lobby at the United Nations Human Rights Committee Meetings in Geneva.
1994	December 15: Ministry of Home Affairs notifies Tanaka and Fukukita that the use of the terms *chōnan* (eldest son) *chōjo* (eldest daughter) will no longer be used on the resident register.
1995	March 1: Change of notation so that all children to be listed by gender and as child (*ko*) using the term formerly used only for out-of-wedlock births, ending the ranking of birth order still in effect on the *koseki* at the time. March 22: Dismissal of the lawsuit against Musashino City and the mayor due to absence of discriminatory terms, "claim lacks benefit."
1999	August 29: Group name changed to Association to Eradicate discrimination in the *koseki* and against children born out of wedlock. November 22: Tanaka-san and her group file lawsuit claiming discrimination in the *koseki* in the *tsuzukigara* (birth order column).
2004	November 1: Ministry of Justice revises rules for notation in the *tsuzukigara* on *koseki* for children born out of wedlock. All children regardless of legitimacy status registered in the eldest-daughter, eldest-son style, enforced immediately for all new registrations.
2005	November 18: Supreme Court dismisses final appeal by the *kōryūkai* for damages due to discrimination in the *tsuzukigara* as not applicable for final appeal.

Glossary

bessei 別姓 separate surnames
chakushutsushi 嫡出子 a "legitimate" child
chakushutsu de nai ko 嫡出でない子 an "illegitimate" child
chōjo 長女 eldest daughter
chōnan 長男 eldest son
fūfubessei 夫婦別姓 married couples with separate surnames, used to describe the political movement for the right to maintain separate surnames in marriage, often shortened to *bessei*
ie seido 家制度 household system
hichakushutsushi 非嫡出子 older term for "illegitimate" child
hittōsha 筆頭者 first person named on a *koseki*, formerly called *koshu*
jijitsukon 事実婚 common-law marriage, literally, real marriage
jūminhyō 住民票 resident register
kekkon todoke 結婚届 marriage notification form
kongaishi 婚外子 child born out of wedlock
koseki 戸籍 family register
ninchi 認知 acknowledge or legally recognize
ninchiko-認知子 an acknowledged child (born out of wedlock, but acknowledged by father)
shiseiji 私生児 "illegitimate" child, older expression
shussei (shusshō) todoke 出生届 Birth Notification form
tsuzukigara 続柄 family or lineal relationship, column on the *koseki* and *jūminhyō* for showing this information

Bibliography

Adair, R. (1996) *Courtship, Illegitimacy and Marriage in Early Modern England*, Manchester: Manchester University Press.

Akaishi, C. (2011) "Single Mothers," in K. Fujimura-Fanselow (ed.), *Transforming Japan: How Feminism and Diversity are Making a Difference*, New York: The Feminist Press.

Alarcon, N., Kaplan, C., and Moallem, M. (1999) "Introduction: Between Woman and Nation," in N. Alarcon, C. Kaplan, and M. Moallem (eds), *Between Woman and Nation: Nationalisms, Transnational Feminisms, and the State*, Durham, NC: Duke University Press.

Amos, T. (2014) "Household Registration and the Dismantling of Edo Outcaste Cultures," in D. Chapman and K.J. Krogness (eds), *Japan's Household Registration System and Citizenship: Koseki, Identification and Documentation*, London: Routledge.

Bailey, Alison (2003) "Privilege: Expanding on Marilyn Frye's 'Oppresion'," in L. Heldke and P. O'Connor (eds), *Oppresion, Privilege, and Resistance: Theoretical Perspectives on Racism, Sexism, and Heterosexism*, Boston: McGraw-Hill.

Buckley, S. (1997) *Broken Silence: Voices of Japanese Feminism*, Berkeley: University of California Press.

Butler, J. (1990) *Gender Trouble: Feminism and the Subversion of Identity*, London: Routledge.

Butler, J. (2004) *Undoing Gender*, New York: Routledge.

CEDAW Draft, Sixty-third Session Report, Advance Unedited Version, July 30, 2015.

Chapman, D. (2008) *Zanichi Korean Identity and Ethnicity*, London: Routledge.

Chapman, D. (2011) "Geographies of Self and Other: Mapping Japan through the *Koseki*," *Asia-Pacific Journal*, 9(29). Online: http://japanfocus.org/-David-Chapman/3565 (accessed May 31, 2013).

Chapman, D. (2012) "No More 'Aliens': Managing the Familiar and the Unfamiliar in Japan," *Asia-Pacific Journal*, 10(40).

Chapman, D. (2014) "Managing 'Strangers' and 'Undecidables': Population Registration in Meiji Japan," in D. Chapman and K.J. Krogness (eds), *Japan's Household Registration System and Citizenship: Koseki, Identification and Documentation*, London: Routledge.

Chapman, D. and Krogness, K.J. (2014) "The Koseki," in D. Chapman and K.J. Krogness (eds), *Japan's Household Registration System and Citizenship: Koseki, Identification and Documentation*, London: Routledge.

Chunn, D. and Lacombe, D. (2000) *Law as a Gendering Practice*, Don Mills, Ontario: Oxford University Press.

Clodd, E. (1921) *Magic in Names and in Other Things*, New York: E.P. Dutton and Company.

Cornell, L.L. and Hayami A. (1986) "The *shūmon aratame chō*: Japan's Population Registers," *Journal of Family History*, 11(4): 311–28.

Dower, J. (1999) *Embracing Defeat: Japan in the Wake of World War II*, New York: W.W. Norton & Company, Inc.

Edwards, W. (1989) *Modern Japan through its Weddings: Gender, Person, and Society in Ritual Portrayals*, Stanford: Stanford University Press.

Emens, E.F. (2007) "Changing Name Changing: Framing Rules and the Future of Marital Names," *The University of Chicago Law Review*, 74(3): 761–863.

Frazer, James George (1922) *The Golden Bough: A Study in Magic and Religion*, London: Macmillan.

Friedman, Marilyn (ed.) (2005) *Women and Citizenship*, New York: Oxford University Press.

Fuess, H. (2004) *Divorce in Japan*, Stanford: Stanford University Press.

Fukushima, Mizuho (1992) *Kekkon to Kazoku: atarashi kankei ni mukete* [Marriage and Family: Towards New Relationships], Tokyo: Iwanami Shoten.

Gelb, J. (2003) *Gender Policies in Japan and the United States: Comparing Women's Movements, Rights, and Politics*, New York: Palgrave Macmillan.

Germer, A. Mackie, V., and Wohr, U. (2014) *Gender, Nation and State in Modern Japan*, London: Routledge.

Gordon, Beate Shirota (1997) *The Only Woman in the Room*, Chicago: University of Chicago Press.

Hayakawa, N. (1995) "Feminism and Nationalism in Japan: 1868–1945," *Journal of Women's History*, 9(4): 108–19.

Hertog, E. (2009) *Tough Choices: Bearing an Illegitimate Child in Contemporary Japan*, Stanford: Stanford University Press.

Highley, Stephen Richard (2000) "Privilege, Power, and Place: The Geography of the American Upper Class," in Tracy E. Ore (ed.), *The Social Construction of Difference and Inequality: Race, Class, Gender, and Sexuality*, New York, NY: Mayfield Publishing.

Hozumi, N. (1940) *Ancestor Worship and Japanese Law*, Tokyo: Hokuseido Press.

Inami, K., Yamazaki, M., and Sano, T. (2008) *Wakariyasui koseki no mikata, yomikata, torikata* [Reading, Observing, and Understanding the Koseki Made Easy], Tokyo: Horei Publishing.

Inoue, H. (1986) *Onna no namae o kaeshite: fūfubessei no susume* [Give Back Women's Names: A Recommendation for Separate Surnames], Osaka: Sōgensha.

Kamoto, I. (2014) "Creating Spatial Hierarchies: The *Koseki*, Early International Marriage and Intermarriage," in D. Chapman and K.J. Krogness (eds), *Japan's Household Registration System and Citizenship: Koseki, Identification and Documentation*, London: Routledge.

Kano, Ayako (2005) "Toward a Critique of Transhistorical Femininity," in B. Molony and K. Uno (eds), *Gendering Modern Japanese History*, Cambridge, MA: Harvard University Press.

Karras, Ruth Mazo (2012) *Unmarriages: Women, Men, and Sexual Unions in the Middle Ages*, Philadelphia: University of Pennsylvania Press.

Kashiwazaki, Chikako (2000) "The Politics of Legal Status: The Equation of Nationality with Ethnonational Identity," in S. Ryang (ed.), *Koreans in Japan: Critical Voices from the Margin*, London: Routledge.

Kawano, S., Roberts, G.S., and Long, S.O. (2014) "Introduction: Differentiation and Uncertainty," in S. Kawano, G.S. Roberts, and S.O. Long (eds), *Capturing Contemporary Japan: Differentiation and Uncertainty*, Honolulu: University of Hawai'i Press.

Kawashima, Y. (1995) "Female Workers: An Overview of Past and Current Trends," in K. Fujimura-Fanselow and A. Kameda (eds), *Japanese Women: New Feminist Perspectives on the Past, Present and Future*, New York: The Feminist Press.

Koseki Jitsumu Kenkyukai [Household Register Guidelines Research Group] (ed.) (2011) *Shoninsha no tame no Koseki jitsumu no tebiki* [Handbook for Koseki Administration for Trainees], Tokyo: Nihon Kajo Publishing Company.

Krogness, K.J. (2008) *The Koseki System and "Koseki Consciousness": An Exploration of the Development and Functions of the Modern Japanese Household Registration System and How it Influences Social Life*, Ph.D. thesis, Copenhagen University.

Krogness, K.J. (2011) "The Ideal, the Deficient, and the Illogical Family: An Initial Typology of Administrative Household Units," in A. Alexy and R. Ronald (eds), *Home and Family in Japan: Continuity and Transformation*, London: Routledge.

Krogness, K.J. (2013) " 'What do you think registry is?': Popular Perceptions of Household Registration in 1950s and 2000s Japan," in I. About, J. Brown, and G. Lonergan (eds), *People, Papers, and Practices: Identification and Registration in Transnational Perspective, 1500–2010*, New York: Palgrave Macmillan.

Krogness, K.J. (2014) "*Jus koseki*: Household Registration and Japanese Citizenship," in D. Chapman and K.J. Krogness (eds), *Japan's Household Registration System and Citizenship: Koseki, Identification and Documentation*, London: Routledge.

Kurotani, S. (2014) "Working Women of the Bubble Generation," in S. Kawano, G.S. Roberts, and S.O. Long (eds), *Capturing Contemporary Japan: Differentiation and Uncertainty*, Honolulu: University of Hawai'i Press.

Lebra, T.S. (1993) *Above the Clouds: Status Culture of the Modern Japanese Nobility*, Berkeley: University of California Press.

Lie, J. (2008) *Zanichi (Koreans in Japan): Diasporic Nationalism and Postcolonial Identity*, Berkeley: University of California Press.

McIntosh, P. (1988) "White Privilege and Male Privilege: A Personal Account of Coming to See Correspondences through Work in Women's Studies (1988)," in M.L. Andersen and P.H. Collins (eds), *Race, Class and Gender: An Anthology*, Belmont: Wadsworth.

Mackie, V. (2003) *Feminism in Modern Japan: Citizenship, Embodiment, and Sexuality*, Cambridge: Cambridge University Press.

Mackie, V. (2014) "Birth Registration and the Right to Have Rights: The Changing Family and the Unchanging *Koseki*," in D. Chapman and K.J. Krogness (eds), *Japan's Household Registration System and Citizenship: Koseki, Identification and Documentation*, London: Routledge.

McLelland, M. (2000) *Male Homosexuality in Modern Japan: Cultural Myths and Social Realities*, Richmond: Curzon.

McLelland, M. (2005) *Queer Japan from the Pacific War to the Internet Age*, Lanham: Rowman & Littlefield.

Maree, C. (2004) "Same-Sex Partnerships in Japan: Bypasses and Other Alternatives," *Women's Studies*, 33: 541–9.

Maree, C. (2014) "Sexual Citizenship at the Intersections of Patriarchy and Heteronormativity: Same Sex Partnerships and the *Koseki*," in D. Chapman and K.J. Krogness (eds), *Japan's Household Registration System and Citizenship: Koseki, Identification and Documentation*, London: Routledge.

Million, Joelle (2003) *Woman's Voice, Woman's Place: Lucy Stone and the Birth of the Woman's Rights Movement*, Westport: Praeger.

Ministry of Health, Labour and Welfare (2006) *Marriage Statistics: Specified Report of Vital Statistics* [Konnin ni kan suru tōkei: jinkōdōtai tōkei tokushuhōkoku: kōseirōdōshō daijin kanbo tōkei jōhōbu], Tokyo: Ministry of Health, Labour and Welfare.

Miyamoto, Y., Ninomiya, S., and Shin, K. (2011) "The Family, Koseki, and the Individual: Japanese and Korean Experiences," *Asia-Pacific Journal: Japan Focus*, 9(36).

Mizuno, N. (2007) "Family and the Family Law in Japan," in M. Tsujimura and E. Yano (eds), *Gender and the Law in Japan*, Sendai: Tohoku University Press.

Molony, B. and Uno, K. (2005) "Introduction," in B. Molony and K. Uno (eds), *Gendering Modern Japanese History*, Cambridge, MA: Harvard University Asia Center.

Mori, K. (2014) "The Development of the Modern Koseki," in D. Chapman and K.J. Krogness (eds), *Japan's Household Registration System and Citizenship: Koseki, Identification and Documentation*, London: Routledge.

Natsuo, Kirino (2003) *OUT*, trans. Stephen Snyder, Tokyo: Kodansha International (originally published in Japanse in 1997).

Neary, I. (2010) *The Buraku Issue and Modern Japan: The Career of Matsumoto Jiichirō*, London: Routledge.

Nemoto, K. (2016) *Too Few Women at the Top: The Persistence of Inequality in Japan*, Ithaca, NY: Cornell University Press.

Ninomiya, S. (2004) "Koseki no tsuzukigara kisai wa hitsuyō ka (3)" [Are the entries in koseki's relationship to parents' column necessary], *Koseki Jihō*, 7(571): 69–83.

Ninomiya, S. (2014) "The Koseki and Legal Gender Change," in D. Chapman and K.J. Krogness (eds), *Japan's Household Registration System and Citizenship: Koseki, Identification and Documentation*, London: Routledge.

Oleksy, E. (2009) "Citizenship Revisited," in E. Oleksy (ed.), *Intimate Citizenships: Gender, Sexualities, Politics*, New York: Routledge.

Ortner, S. (1974) "Is Female to Male as Nature is to Culture," in M. Rosaldo and L. Lamphere (eds), *Woman, Culture and Society*, Stanford: Stanford University Press.

Parrenas, R.S. (2001) "Transgressing the Nation-State: The Partial Citizenship and 'Imagined (Global) Community' of Migrant Filipina Domestic Workers," *Signs*, 26(4): 1129–54.

Pateman, C. (1988) *The Sexual Contract*, Stanford: Stanford University Press.

Plutschow, H. (1995) *Japan's Name Culture: The Significance of Names in a Religious, Political and Social Context*, Kent: Japan Library.

Rich, Adrienne (1979) "Compulsory Sexuality and Lesbian Existence," in A. Rich, *On Lies, Secrets, and Silence: Selected Prose 1966–1978*, New York: W.W. Norton.

Roberts, Glenda S. (2014) "Work and Life in Challenging Times: A Kansai Family across the Generations," in S. Kawano, G.S. Roberts, and S.O. Long (eds), *Capturing Contemporary Japan: Differentiation and Uncertainty*, Honolulu: University of Hawai'i Press.

Ryang, Sonia (2000) *Koreans in Japan: Critical Voices from the Margin*, London: Routledge.

Sakakibara, Fujiko (1992) *Josei to Koseki: Fūfubesseijidai ni mukete* [Women and the Koseki: Towards the Era for Separate Surname Couples], Tokyo: Akashi Shoten.

Satō, Bunmei (1996) "Shōchō tennōsei ni totte, koseki to wa nanika" [What is the Koseki for a Symbolic Emperor System?], in Tennōsei Kenkyukai (ed.), *Koseki Kaitai Koza*, Tokyo: Shakai Hyōronsha.

Seidensticker, E. (1976) *The Tale of Genji* (by Murasaki Shikibu), Tokyo: Charles E. Tuttle Company.

Shimazu, Y. (1994) "Unmarried Mothers and Their Children in Japan," *US Japan Women's Journal, English Supplement*, 6: 83–110.

Smith, Robert J. (1974) *Ancestor Worship in Contemporary Japan*, Stanford: Stanford University Press.

Smith, Robert J. (1987) "Gender Inequality in Contemporary Japan," *Journal of Japanese Studies*, 13(1): 1–25.

Smith, R. and Wiswell, E.L. (1982) *The Women of Suye Mura*, Chicago: University of Chicago Press.

Stannard, U. (1977) *Mrs. Man*, San Francisco: Germainbooks.

Steinhoff, P.G. (2014) "Introduction," in *Going to Court to Change Japan: Social Movements and the Law in Contemporary Japan*, Ann Arbor: Center for Japanese Studies, University of Michigan Press.

Stone, Linda (1997) *Kinship and Gender: An Introduction*, Boulder: Westview Press.

Suzuki, Nobue (2010) "Outlawed Children: Japanese Filipino Children, Legal Defiance and Ambivalent Citizenships," *Pacific Affairs*, 83(1): 31–50.

Tanaka, K. (1995) "The New Feminist Movement in Japan, 1970–1990," in K. Fujimara-Fanselow and A. Kameda (eds), *Japanese Women: New Feminist Perspectives on the Past, the Present, and Future*, New York: The Feminist Press.

Tanaka, K. (2012) "Surnames and Gender in Japan: Women's Challenges in Seeking Own Identity," *Journal of Family History*, 37(2): 232–40.

Tanaka, Sumiko (2004a) Nakusō Koseki to Kongaishi Sabetsu Koryūkai (eds), *Nakusō Kongaishi Josei e no Sabetsu* [Let's End Discrimination Against Women and Children Born Out of Wedlock], Tokyo: Akashi Shoten.

Tanaka, Sumiko (2004b) "Interview: Fifteen Years of Fighting Against the Discrimination of Children Born Out of Wedlock," *Women's Asia 21: Voices from Japan*, 13.

Toshitani, Nobuyoshi (.1994). "The Reform of Japanese Family Law and Changes in the Family System,". *U.S. Japan Women's Journal, English Supplement*, 6.

Tsurumi, Patricia (1992) *Factory Girls: Women in the Thread Mills of Meiji Japan*, Princeton: Princeton University Press.

Ueno, Chizuko (1987) "The Position of Japanese Women Reconsidered," *Current Anthropology*, 28(4): 75–84.

Ueno, Chizuko (2009) "The Trap of Separate Surnames for Married Couples," in *The Modern Family in Japan: Its Rise and Fall*, (p.199-205) Victoria, Australia: Trans Pacific Press.

Wakakuwa, M. and Fujimura-Fanselow, K. (2011) "Backlash Against Gender Equality after 2000," in K. Fujimura-Fanselow (ed.), *Transforming Japan: How Feminism and Diversity are Making a Difference*, New York: The Feminist Press.

Warner, M. (1993) *Fear of a Queer Planet: Queer Politics and Social Theory*, Minneapolis: University of Minnesota Press.

Watanabe, Y. (1963) (assisted by Max Rheinstein) "The Family and the Law: The Individualistic Premise and Modern Japanese Family Law," in (ed) Arthur Taylor von Mehren (ed.), *Law in Japan: The Legal Order in a Changing Society*, Cambridge, MA: Harvard University Press.

White, Linda (2014) "Challenging the Heteronormative Family in the *Koseki*: Surname, Legitimacy, and Unmarried Mothers," in D. Chapman and K.J. Krogness (eds.), *Japan's Household Registration System and Citizenship: Koseki, Identification and Documentation*, London: Routledge.

Wiesner, C. (2013) "Women's Partial Citizenship: Cleavages and Conflicts Concerning Women's Citizenship in Theory and Practice," in J.M. Rosales, T. Pulkkinen, and K. Palonen (eds.), *The Ashgate Research Companion to the Politics of Democratization in Europe*, London: Ashgate.

Wildman, S.M. (2000) "Ending Male Privilege: Beyond the Reasonable Woman," *Michigan Law Review*, 98.

Yamaguchi, Tomomi (2014) " 'Gender-Free' Feminism in Japan: A Story of Mainstreaming and Backlash," *Feminist Studies*, 40(3): (541–572).

Yuval-Davis, N. (1993) "Gender and Nation," *Ethnic and Racial Studies*, 16(4): 621–32.

Index

Page numbers in *italic* refer to figures

For Product Safety Concerns and Information please contact our EU representative GPSR@taylorandfrancis.com Taylor & Francis Verlag GmbH, Kaufingerstraße 24, 80331 München, Germany